Series A Vol. 56

Series Editor: Leon O. Chua

STRANGE NONCHAOTIC ATTRACTORS

Dynamics between Order and Chaos in Quasiperiodically Forced Systems

Ulrike Feudel
University of Oldenburg, Germany

Sergey Kuznetsov
Saratov State University, Russia

Arkady Pikovsky
University of Potsdam, Germany

NEW JERSEY • LONDON • SINGAPORE • BEIJING • SHANGHAI • HONG KONG • TAIPEI • CHENNAI

Published by

World Scientific Publishing Co. Pte. Ltd.
5 Toh Tuck Link, Singapore 596224
USA office: 27 Warren Street, Suite 401-402, Hackensack, NJ 07601
UK office: 57 Shelton Street, Covent Garden, London WC2H 9HE

British Library Cataloguing-in-Publication Data
A catalogue record for this book is available from the British Library.

STRANGE NONCHAOTIC ATTRACTORS

ISBN-13 978-981-256-633-1
ISBN-10 981-256-633-3

Printed in Singapore

STRANGE NONCHAOTIC ATTRACTORS

Dynamics between Order and Chaos in Quasiperiodically Forced Systems

To Fred and Frank (UF)

To memory of my parents (SK)

To Maria (AP)

Preface

This book is devoted to strange nonchaotic attractors – dynamical states that appear in quasiperiodically forced systems and lie in between chaos and order. They have been discovered in a seminal paper by Grebogi, Ott, Pelikan, and Yorke [1984]. Since then these objects have been observed experimentally and have been found in different models. Although a complete mathematical description of SNAs is still to be done, many of their properties are well understood on a physical level, and this is the subject of our book.

We try to describe here strange nonchaotic attractors on a level accessible for graduate students having a basic knowledge of nonlinear dynamics. We show how different tools of nonlinear dynamics, like bifurcation analysis and Lyapunov exponents, work for these systems. Although our presentation mainly relies on our works during the last decade, we also have tried to review many papers appearing in this field.

During these years we had the pleasure to discuss various aspects of the dynamics of strange nonchaotic attractors with our colleagues and co-authors, to whom we express our sincere gratitude: V. Afraimovich, V. Anischshenko, B. Bezruchko, L. Bunimovich, S. Datta, M. Ding, P. Glendinning, C. Grebogi, B. Hunt, N. Ivan'kov, K. Kaneko, G. Keller, J. Kurths, A. Kuznetsov, Y.-C. Lai, Yu. Maistrenko, E. Neumann, T. Nishikawa, A. Osbaldestin, V. Oseledets, H. Osinga, E. Ott, A. Politi, R. Ramaswamy, M. Rosenblum, Ye. Seleznev, M. Shrimali, Ya. Sinai, O. Sosnovtseva, S. Sinha, J. Stark, R. Sturman, I. Sushko, J. Wiersig, A. Witt, H. L. Yang, M. Zaks.

A large part of our research on strange nonchaotic attractors has been performed in the Nonlinear Dynamics Group at the University of Potsdam, Germany. It is our great pleasure to thank the leader of the group Jürgen

Kurths, as well as all colleagues who ensured a creative atmosphere there. Our especial thanks go to Birgit Voigt and Jörg-Uwe Tessmer for their technical support.

Ulrike Feudel, Sergey Kuznetsov, Arkady Pikovsky

Contents

Preface vii

1. Introduction 1

1.1 Periodicity and quasiperiodicity 1
1.2 Robustness of quasiperiodic motions 3
1.3 Strange nonchaotic attractors 5
1.4 What is in the book 6

2. Models 9

2.1 Differential equations and maps 9
2.2 Quasiperiodically forced one-dimensional maps 11
2.2.1 GOPY model (modulated pitchfork map) 12
2.2.2 Forced circle map 14
2.2.3 Skew shift 15
2.2.4 Forced logistic map 16
2.2.5 Harper model 20
2.3 Quasiperiodically forced high-dimensional maps 20
2.4 Quasiperiodically forced continuous-time systems 21
2.4.1 Forced overdamped pendulum 22
2.4.2 Forced Duffing oscillator 23
2.5 Experiments 26
2.6 Bibliographic notes 27

3. Rational approximations 29

3.1 Properties of rational approximations of irrationals 30
3.2 Rational approximations to quasiperiodic forcing 32

3.3 Checking strangeness of SNA through rational approximations 33
3.3.1 Rational approximations to a smooth attractor . . . 33
3.3.2 Rational approximations to an SNA: An example . . 33
3.3.3 Rational approximations to an SNA: General consideration 36
3.3.4 Different examples 39
3.4 Bibliographic notes 42

4. Stability and Instability 45

4.1 Theoretical consideration 45
4.2 Numerical examples 50
4.2.1 Discrete time mappings 50
4.2.2 Continuous time systems 54
4.3 Bibliographic notes 55

5. Fractal and statistical properties 57

5.1 Fractal properties of SNA 57
5.2 Correlations and spectra of SNA 60
5.2.1 Power spectra of regular and irregular motions . . . 60
5.2.2 Spectral properties of fractal tori 63
5.2.3 Singular continuous spectrum in an SNA 65
5.2.4 Theoretical description of the singular continuous spectrum 69
5.3 Bibliographic notes 73

6. Bifurcations in quasiperiodically forced systems and transitions to SNA 75

6.1 Smooth and non-smooth bifurcations 76
6.2 Bifurcations in the quasiperiodically forced logistic map . . 77
6.2.1 Torus doubling 79
6.2.2 Non-smooth tori collision beyond period-doubling . . 81
6.2.3 Fractalization of the torus 83
6.2.4 Interior crisis 85
6.2.5 Boundary crisis 91
6.2.6 Basin boundary bifurcation 95
6.3 Bifurcations in the quasiperiodically forced circle map . . . 102
6.3.1 Smooth saddle-node bifurcation of tori 104
6.3.2 Non-smooth collision of a stable and an unstable torus 104

6.3.3 Phase-locking regions under quasiperiodic forcing . . 107
6.3.4 Non-smooth pitchfork bifurcation 111
6.4 Loss of transverse stability: blowout transition to SNA . . . 113
6.5 Intermittency 119
6.6 Bibliographic notes 127

7. Renormalization group approach to the onset of SNA in maps with the golden-mean quasiperiodic driving 131

7.1 Introduction: The main idea of the renormalization group analysis 131
7.2 The basic functional equations for the golden-mean renormalization scheme 134
7.3 A review of critical points 136
7.3.1 Classic GM point 137
7.3.2 Critical point of the blowout birth of SNA 137
7.3.3 Critical points of torus doubling terminal and torus collision terminal 139
7.3.4 Critical point of torus fractalization 141
7.4 RG analysis of the classic GM critical point 142
7.5 RG analysis of the blowout birth of SNA 145
7.6 RG analysis of the TDT critical point 154
7.7 RG analysis of the TCT critical point 165
7.8 RG analysis of the TF critical point 178
7.9 Critical behavior in realistic systems 187
7.10 Conclusion 193
7.11 Bibliographic notes 195

Bibliography 197

Index 211

Chapter 1

Introduction

This book is devoted to strange nonchaotic attractors, which are typical to quasiperiodically forced systems. In this introductory chapter we describe what is the place of these objects in nonlinear dynamics, what is common and what is the difference of SNA compared to quasiperiodicity and chaos.

1.1 Periodicity and quasiperiodicity

It is natural to classify possible dynamical regimes observed in nonlinear systems by their complexity. The most simple nontrivial regime is a periodic one. Already the consideration of conservative systems with one degree of freedom leads to such solutions. These solutions in conservative systems appear in families and are not isolated, a small perturbation or a change of initial conditions leads to a transition to another periodic solution, in a nonlinear system the period generally changes, too.

In autonomous dissipative dynamical systems periodic solutions appear when there are mechanisms of energy supply and energy dissipation present in the system. As a result, a periodic motion with a certain amplitude appears as a structurally stable regime: it reestablishes after a perturbation; these solutions are often called self-sustained oscillations. In the phase space of a dynamical system such stable periodic motions are described by limit cycles. The minimal dimension of the phase space for a limit cycle to exist is two; moreover, in two-dimensional systems more complex structurally stable regimes are impossible.

The next complex dynamical regime is quasiperiodicity. This regime appears already in elementary linear conservative systems with two degrees of freedom, where a general solution has the form

$$x(t) = a_1 \cos(\omega_1 t + \varphi_1^0) + a_2 \cos(\omega_2 t + \varphi_2^0) , \tag{1.1}$$

where ω_1 and ω_2 are the natural frequencies depending on the parameters of the system. Solution (1.1) is periodic only if the ratio of the frequencies is rational, i.e. when

$$\frac{\omega_1}{\omega_2} = \frac{p}{q} \tag{1.2}$$

with integers p, q. In this case the period is $T = p\frac{2\pi}{\omega_1} = q\frac{2\pi}{\omega_2}$. By contrast, if the ratio of frequencies is an irrational number, (1.2) cannot be valid for any pair of integers p, q, and the motion is quasiperiodic. More generally, to give a definition of a quasiperiodic function of time let us introduce a function

$$g(\phi_1, \phi_2, \ldots, \phi_n) \tag{1.3}$$

of n arguments, which is 2π-periodic in each of them. These arguments (naturally called phases) are linear functions of time

$$\dot{\phi}_1 = \omega_1 , \quad \ldots \quad \dot{\phi}_n = \omega_n . \tag{1.4}$$

Next, we demand that the frequencies ω_i are linearly independent, i.e. the relation

$$k_1\omega_1 + k_2\omega_2 + \ldots + k_n\omega_n = 0 \tag{1.5}$$

cannot be satisfied for any set of integers $k_1 \ldots k_n$. Then $g(\phi_1, \phi_2, \ldots, \phi_n)$ describes a quasiperiodic motion with n incommensurate frequencies. Expression (1.1) is a particular form of (1.3).

Quasiperiodic solutions of type (1.3) naturally appear in integrable Hamiltonian systems with n degrees of freedom, where a transformation to angle-action variables is possible, and the angle variables obey (1.4). Note, however, that because the action variables may be changed under a perturbation , and the frequencies depend on these variables, condition (1.5) does not survive under a perturbation of initial conditions and/or parameters. In terms of our discrimination of periodic and quasiperiodic motions, this means that under small perturbations a quasiperiodic motion may become periodic, and vice versa. This already makes a study of structural stability of quasiperiodic regimes nontrivial, but before considering it closely we discuss the possibility to observe quasiperiodic motions in dissipative systems.

A natural way to construct a quasiperiodic regime in a dissipative dynamical system is to take a high-dimensional integrable conservative model as described above and to assume that due to supply of energy for small amplitudes (values of action variables) and to dissipation of energy for large amplitudes, a certain stable values of amplitudes is established. The dynamics then is described by the phase variables solely, according to (1.4).

There is also another way to construct a quasiperiodic state in a dissipative system. One starts here with a simple stable steady state. This state can lose stability when parameters of the system are changed, and produce a stable limit cycle via a Hopf bifurcation. Now one can assume that with a further change of parameters the periodic motion can become unstable toward the appearance of another periodic component (mode) with some other frequency, the amplitude of this mode is assumed to be stabilized at some level. As a result of this secondary Hopf bifurcation, also called Neimark-Sacker bifurcation, a quasiperiodic motion with two frequencies can appear. Assuming as a hypothesis that further secondary bifurcations can occur, one can imagine the appearance of quasiperiodic motions of higher order. Exactly this picture has been drawn by Landau in his book on Hydrodynamics [Landau and Lifshitz 1987], where he tried to imagine how a hypothetical way to turbulence via a consecutive complication of the dynamical state can occur, as a parameter (for turbulence a natural parameter is the Reynolds number) changes. Landau was not aware of the possibility of chaotic dynamics, nowadays a transition to turbulence is usually related to the appearance of chaos rather than to a high-dimensional quasiperiodic state.

1.2 Robustness of quasiperiodic motions

We have seen that quasiperiodic motion is defined by a rather subtle characteristics, namely by incommensurate frequencies. One can naturally ask, whether this property is not only mathematically, but also physically relevant, i.e. whether in a real world with inevitable noise and measurement errors one can really distinguish periodic and quasiperiodic regimes. To answer this question it is natural to introduce an observation time and to distinguish periodicity and quasiperiodicity with respect to this time. Considering relation (1.2) one can compare the possible period of the process with this observation time: if the ratio of the frequencies (1.2) is rational but p and q are so large that the period exceeds the observation time, then it is not possible to distinguish this periodic regime from a quasiperiodic one. In other words, practically a complex periodic state may be considered as a quasiperiodic one.

There is also another aspect in the relation between periodic and quasiperiodic regimes, it is related to a parameter dependence. Indeed, if the frequencies ω_1 and ω_2 depend on parameters, then their ratio depends on the parameters too and it might happen that already a small parameter variation leads to transitions periodicity $\leftrightarrow$ quasiperiodicity.

This aspect of robustness of a quasiperiodic state is a complex issue that still is not resolved in its full generality. A rather complete answer exists in the case of two frequencies (1.2) only. The main point is that one has to consider the phases ϕ_i not as independently rotating ones, but as coupled dynamical variables. For two phases one then writes

$$\dot{\phi}_1 = \omega_1 + F_1(\phi_1, \phi_2) , \qquad \dot{\phi}_2 = \omega_2 + F_2(\phi_1, \phi_2) , \tag{1.6}$$

with coupling functions F_1, F_2 which are 2π-periodic in each argument. If these functions are small, the dynamics (1.6) on the two-dimensional torus $0 \le \phi_1, \phi_2 < 2\pi$ can be reduced to a circle map, and as a result a full characterization of periodic and quasiperiodic regimes can be achieved. We will not present the whole theory here (see [Katok and Hasselblatt 1995] for details), we only mention two important features:

(1) Periodic regimes, i.e. regimes where the observed frequencies $\Omega_{1,2} = \langle\dot{\phi}_{1,2}\rangle$ are in a rational relation like (1.2), are structurally stable: they exist for certain parameter ranges (open sets of parameters). Contrary to this, quasiperiodic states with incommensurate $\Omega_{1,2}$ are isolated. This means that e.g. if one parameter is changed, a quasiperiodic regime with a given ratio Ω_1/Ω_2 exist for one particular value of this parameter; if two parameters are varied, this regime exists on a line in the parameter plane , etc. Therefore one can say that periodic regimes are topologically stable and quasiperiodic regimes are not.
(2) For small nonlinearities in (1.6), i.e. for small $F_{1,2}$, the measure of all parameter values for which periodic regimes are observed is small while the measure of quasiperiodic states is large. One can interprete this as the abundance of quasiperiodicity: if parameters are chosen "at random", then a quasiperiodic state will be observed with high probability. For larger nonlinearities, i.e. for large $F_{1,2}$, the portion of periodic regimes grows while that of quasiperiodic regimes decreases.

These two properties mean that although being structurally unstable, quasiperiodic regimes are nevertheless physically observable. However, if one wants to have a quasiperiodic regime with a particular ratio of the frequencies, then special efforts are necessary to adjust the system parameters to the needed values.

The situation with quasiperiodicity is not so clear when the number of frequencies is larger than two. Certainly, in this case quasiperiodic states are structurally unstable as well, i.e. with a small change of parameters another regime, e.g. a periodic one, can be observed. Moreover, because the dimension of the system for phases (1.4) is large now, structurally stable regimes that are more complex than periodic ones are possible. In the famous paper by Ruelle and Takens [1971] (see also an extension in [Newhouse et al. 1978]) it was shown that also structurally stable chaotic states can occur if one adds some particular arbitrary small nonlinear terms to Eqs. (1.4). However, again, like in the simplest case of two frequencies, one can expect that from the probabilistic point of view quasiperiodic regimes are abundant while periodic and chaotic ones are rather unprobable, at least for small nonlinearities. This picture has been confirmed by numerical calculations in [Grebogi et al. 1983b, 1985], where a statistical test on how often randomly chosen functions F_i lead to quasiperiodicity has been performed.

Concluding, we can say that quasiperiodic motions, being structurally unstable, are nevertheless physically observable. However, to observe a quasiperiodic state with a particular ratio of frequencies, special efforts on adjusting the system's parameters must be made. Otherwise, it is possible to fall into periodicity, moreover, periodic states become more and more probable for large nonlinearities (large coupling between modes).

1.3 Strange nonchaotic attractors

In typical classifications of dynamical complexity chaos is considered as the next stage beyond quasiperiodicity. Chaos is usually defined as a dynamical regime with sensitive dependence on initial conditions; quantitatively this is characterized by a positive largest Lyapunov exponent. Contrary to this, in a quasiperiodic regime the largest Lyapunov exponents are equal to zero, their number is the number of phase variables in (1.4). Chaos also shares many statistical properties with noise, in particular it possesses decaying correlations. In the phase space of a dissipative system the object corresponding to chaos is a strange attractor, which is a fractal set of zero measure.

A ***strange nonchaotic attractor (SNA)*** is an object that lies in between quasiperiodicity and chaos. This concept has been introduced in a seminal paper by Grebogi, Ott, Pelikan, and Yorke [1984], and is related to a special class of dissipative systems with a quasiperiodic forcing of type (1.3). The force is characterized by n phases $\phi_1 \dots \phi_n$, and the driven system by some variables $x_1 \dots x_m$.

The term ***strange*** means that the dependence of the dynamical variables on the phases is not given by smooth relations, but constitutes some fractal. This is in contrast to a possible quasiperiodic dynamics of $x_l(t)$, where these variables are smooth functions of the phases ϕ_i and thus are described by functions of type (1.3).

The term ***nonchaotic*** means the absence of sensitive dependence on initial conditions. Quantitatively, the largest Lyapunov exponent corresponding to the variables $x_1 \dots x_m$ is negative.

A possibly complete description of strange nonchaotic attractors is the subject of this book. Numerous studies have shown that SNA is not a degenerate object existing for some special parameter values, but can be observed generally in quasiperiodically forced systems. Moreover, in such systems it is typically found in the transition region from order to chaos. One can say that in such a transition "strangeness" appears prior to "chaoticity". Moreover, there is a class of systems where chaos is simply impossible because the phase space has a too low dimension (e.g. if the dynamical variable x is just one-dimensional). In this case an SNA provides, possibly, the maximal complexity in the dynamics.

1.4 What is in the book

Strange nonchaotic attractors have been extensively studied in physical literature, with approximate analytic methods, numerically, and experimentally. However, rigorous mathematical works are rather rare. In this book we follow mainly analytic and numerical studies and try to characterize SNAs from different sides. We start in Chapter 2 with a description of different dynamical models leading to SNAs. Here we also describe experiments with SNA and discuss relations to other physical and dynamical problems. In the following chapters these models will be used to illustrate different features of SNAs: How one can describe SNA using periodic approximations to quasiperiodic forcing (Chapter 3); how one can characterize the stability of motion on SNA (Chapter 4); what are properties of correlations and spectra of SNA (Chapter 5). In Chapter 6 we analyze different transitions from regular to strange nonchaotic behavior and further to chaos. In many cases these transitions demonstrate remarkable properties of self-similarity that can be explained with a renormalization group approach (Chapter 7).

Chapter 2

Models

In this chapter we introduce most the prominent nonlinear models where SNA can be found, they are used in the subsequent chapters to discuss the properties of SNAs. Here we illustrate different dynamical regimes with pictures of the phase portraits and some elementary characterization of the dynamics. A more detailed consideration of some representative models will be done in the next chapters, where we introduce advanced methods of analysis and demonstrate them using these models.

2.1 Differential equations and maps

First we discuss the relation between quasiperiodically forced continuous-time systems and maps. For simplicity, we restrict our consideration to a two-frequency forcing. A general continuous-time dynamical system with such a force can be written as

$$\frac{d\mathbf{x}}{dt} = \mathbf{F}(\mathbf{x}, \phi_1, \phi_2) \; , \tag{2.1}$$

$$\frac{d\phi_1}{dt} = \omega_1 \; , \tag{2.2}$$

$$\frac{d\phi_2}{dt} = \omega_2 \; . \tag{2.3}$$

Here $\mathbf{x} = \{x_1, \ldots, x_m\}$ is the state vector of the driven system, and the functions $\mathbf{F} = \{f_1, \ldots, f_m\}$ are 2π-periodic in the arguments ϕ_1, ϕ_2. To accomplish a reduction of continuous-time system (2.1)-(2.3) to a map, we use a stroboscopic method. Let us observe the system at the moments of time where the phase $\phi_1(t)$ attains some prescribed value. The time interval between these time instants is $T_1 = \frac{2\pi}{\omega_1}$, these instants $t_n = nT_1$

can be labeled with discrete time index n. The values of the phase ϕ_2 at these moments of time are

$$\phi_2(t_n) = 2\pi \frac{\omega_2}{\omega_1} n + \phi_2(0) \tag{2.4}$$

Introducing a new phase normalized to the interval $(0,1)$ as $\theta_n = (2\pi)^{-1}\phi_2(t_n)$, we can write for this phase the recursion $\theta_{n+1} = \theta_n + w$, where $w = \frac{\omega_2}{\omega_1}$. Finally, the evolution over the time interval T_1 according to (2.1) defines a mapping $\mathbf{x}(nT_1) \to \mathbf{x}((n+1)T_1)$ which depends on the value of the second phase $\phi_2(nT_1)$ (the dependence on the value of the first phase is not important, as it is the same for all chosen time instants t_n). This means that we can reduce (2.1)-(2.3) to a mapping

$$\mathbf{x}_{n+1} = \mathbf{f}(\mathbf{x}_n, 2\pi\theta_n)\ , \tag{2.5}$$

$$\theta_{n+1} = \theta_n + w \pmod 1\ . \tag{2.6}$$

We have included the operation $\pmod 1$ in the equation for the discrete phase θ to underline its periodicity: the values θ and $\theta + 1$ are equivalent.

The main parameter that defines the properties of the forcing in (2.5, 2.6) is the ratio of two frequencies w. If this ratio is rational $w = \frac{p}{q}$, then $\theta_q = \theta_0$ and the forcing is periodic. In this case one can consider the q-th iteration of (2.5, 2.6), this iteration will be an autonomous mapping. If w is an irrational number, then the forcing is quasiperiodic. Usually one takes for w the "simplest" irrational number – the inverse of the golden mean

$$\omega = \frac{\sqrt{5}-1}{2}\ . \tag{2.7}$$

If not specially stated, this ratio of frequencies will be used in different examples in this book. This choice is not only convenient for a theoretical analysis (see Chapters 3 and 7 below), but it provides also a better visibility of fine structures in the state and in the parameter space – and the investigation of these structures is often the most interesting challenge!

Equations (2.5, 2.6) constitute a general discrete model of a quasiperiodically two-frequency forced nonlinear system. For a larger number of driving frequencies one should straightforwardly extend (2.5, 2.6) by adding further discrete phases.

The connection between the models (2.1)-(2.3) and (2.5, 2.6) can be used in two ways. On one hand, the reduction of (2.1)-(2.3) to (2.5, 2.6) as described above can be used for the analysis of a continuous-time dynamics, e.g. for the visualization of the phase portraits. On the other hand, one can study different properties of quasiperiodically forced systems on models of type (2.5, 2.6), which are sometimes simpler to handle, but nevertheless one can generalize the conclusions to continuous-time systems. However, the last statement is only true if $\mathbf{f}$ is invertible. Otherwise such models cannot be directly related to any continuous-time system of type (2.1)-(2.3).

2.2 Quasiperiodically forced one-dimensional maps

One-dimensional maps are simple objects that demonstrate a rich dynamics, including chaos. They provide also the simplest examples of strange nonchaotic attractors. A general form of a quasiperiodically forced one-dimensional map is

$$x_{n+1} = f(x_n, 2\pi\theta_n) , \tag{2.8}$$

$$\theta_{n+1} = \theta_n + w \pmod 1 . \tag{2.9}$$

It is rather easy to characterize such systems in terms "chaotic" – "not chaotic", as the calculation of the largest Lyapunov exponent is here straightforward:

$$\lambda = \left\langle \ln \left| \frac{df}{dx} \right| \right\rangle . \tag{2.10}$$

Averaging in (2.10) includes also averaging over θ which can be easily performed using the observation that θ is uniformly distributed on the interval $0 \leq \theta < 1$.

There is a simple argument that if the function f is monotonous in x, then the Lyapunov exponent of an attractor cannot be positive, i.e. the motion is non-chaotic. Indeed, a positive Lyapunov exponent would mean that the attractor is expanding in x-direction, and due to monotonicity of the mapping this expansion cannot be bounded. Thus a bounded attractor with a positive Lyapunov exponent is impossible. Such an attractor can, of course, occur if the mapping is non-monotonous, like e.g. the logistic map, as here the existence of a bounded attractor can be ensured due to folding. From this argument it follows, that for monotonous one-dimensional mappings which possess a bounded attractor only its strangeness has to be tested to show the existence of SNA. For more general mappings we have additionally to check non-chaoticity by calculating the Lyapunov exponent according to (2.10).

2.2.1 *GOPY model (modulated pitchfork map)*

We start with the historically first example of SNA introduced by Grebogi, Ott, Pelikan, and Yorke [1984], we will call it GOPY-map in the subsequent chapters. To introduce it let us take a simple autonomous monotonous one-dimensional map $x_{n+1} = a\tanh(x_n)$. This symmetric map has a stable fixed point $x = 0$ if $|a| < 1$. For $|a| > 1$ this fixed point becomes unstable, and the attractor is a period-2 cycle for $a < -1$, while there are two symmetric stable fixed points for $a > 1$. Thus, this mapping demonstrates either a period-doubling or a pitchfork bifurcation, depending on the sign of a. Notice that $x = 0$ is a solution for all values of a.

In [Grebogi et al. 1984] this map was modified by introducing a quasiperiodic modulation of parameter a in the following way, leading to the famous GOPY model:

$$x_{n+1} = 2a\tanh(x_n)\cos(2\pi\theta_n) \ , \tag{2.11}$$

$$\theta_{n+1} = \theta_n + \omega \ . \tag{2.12}$$

One can see, that the line $x = 0 \quad 0 \leq \theta \leq 1$ is a solution for all values of a. To investigate the stability of this solution, we have to calculate the average Lyapunov exponent λ_0 on it (it is not the Lyapunov exponent of the whole system, because we do not know whether $x = 0$ is an attractor). For small x one writes $\tanh x \approx x$ and

$$\ln\left|\frac{x_N}{x_0}\right| = \sum_{n=0}^{N-1} \ln|2a\cos(2\pi\theta_n)| \ .$$

Thus

$$\lambda_0 = \lim_{N\to\infty}\frac{1}{N}\ln\left|\frac{x_N}{x_0}\right| = \langle \ln|2a\cos(2\pi\theta)|\rangle \ . \tag{2.13}$$

Using the uniform distribution of phases θ we get finally

$$\lambda_0 = \int_0^1 d\theta\ \ln|2a\cos(2\pi\theta)| = \ln|a|. \tag{2.14}$$

Thus for $|a| < 1$ the trivial solution $x = 0$ is stable (and is in fact the global attractor of the system), and for $|a| > 1$ some other attractor with $x \neq 0$ appears. It is clear that this attractor lies in the strip $|x| < 2a$. The nontrivial character of this attractor follows from the following observation. Because the multiplicative factor $\cos(2\pi\theta)$ vanishes for $\theta_0 = \frac{1}{4}$, then for $\theta_1 = \theta_0 + \omega$ the value of x is zero. Because the image of zero remains zero under (2.11), x vanishes also for the whole trajectory of θ_0, i.e. for all points $\frac{1}{4} + n\omega$ (mod 1). But these points are dense on the interval $0 \leq \theta < 1$. This means that although the attractor of (2.11, 2.12) lies in some strip outside $x = 0$, it is pinched in a dense number of points. The picture of the attractor is shown in Fig. 2.1. This attractor can be characterized in different ways, many of these approaches will be discussed in subsequent chapters. We mention here only, that the modulation of parameter a in (2.11, 2.12) can be considered as leading to a "mixture" of pitchfork and period-doubling bifurcations, and this mixture yields the appearance of SNA in this example.

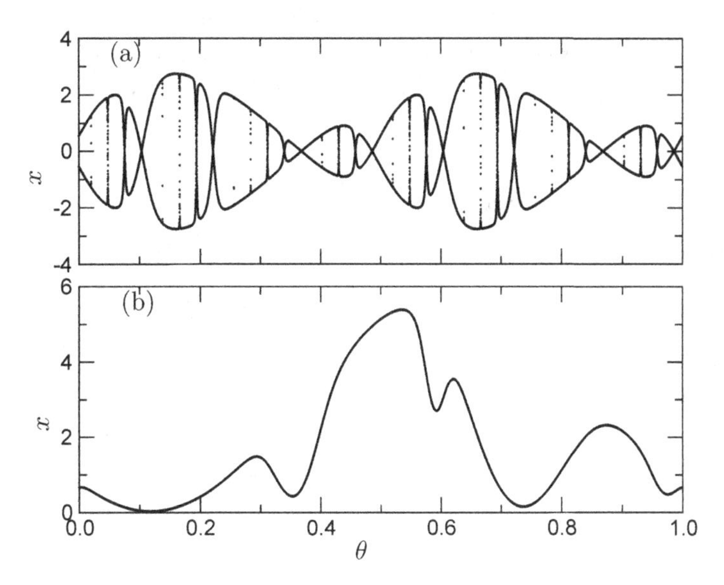

Fig. 2.1 Panel (a) shows the SNA in model (2.11, 2.12) for $a = 1.5$. 50000 points allow one to see many pinches, but still one cannot see that they are dense on the attractor – for this one needs to plot many more points. In panel (b) we show for comparison a smooth quasiperiodic attractor in model (2.17, 2.18) for $a = 1.5$, $b = 1.01$.

We discuss next two modifications of model (2.11, 2.12) where additional terms in the mapping and additional parameters are included. Already in the pioneering paper [Grebogi, Ott, Pelikan, and Yorke 1984] the following modification has been introduced

$$x_{n+1} = 2a\tanh(x_n)\cos(2\pi\theta_n) + \alpha\cos(2\pi(\theta_n + \beta)) , \tag{2.15}$$

$$\theta_{n+1} = \theta_n + \omega . \tag{2.16}$$

Now the trivial state $x = 0$ is no more a solution, and the analysis becomes rather complex. A detailed study shows that for $\alpha \neq 0$ the attractor in system (2.15, 2.16) is not strange (see Fig. 4.2 below and cf. Chapter 3).

Another modification of (2.11, 2.12) has been suggested by Glendinning [2004], in this version a constant is added to the modulation term:

$$x_{n+1} = 2a \tanh(x_n)(b + \cos(2\pi\theta_n)) \,, \tag{2.17}$$

$$\theta_{n+1} = \theta_n + \omega \,. \tag{2.18}$$

Here $x = 0$ remains a solution on the whole plane of parameters (a, b). However, the modulation term $b + \cos(2\pi\theta_n)$ does not change sign if $b > 1$. For these values of parameters one observes not a "mixture" of the pitchfork and period-doubling bifurcations, but only the pitchfork bifurcation which leads to a smooth attractor like that shown in Fig. 2.1b, see also Fig. 7.11 below.

2.2.2 *Forced circle map*

The autonomous circle map

$$\varphi_{n+1} = \varphi_n + f(2\pi\varphi_n) \pmod 1 \tag{2.19}$$

is a popular model in nonlinear dynamics. The variable φ is defined on a circle $0 \leq \varphi < 1$ and the function f is 2π-periodic. If $2\pi|f'| < 1$, the circle map is monotonous and possible regimes described by (2.19) are periodic orbits and quasiperiodic motions. In the periodic case the attractor consists typically of a finite number of points on the circle. These points correspond to a stable periodic orbit. In the quasiperiodic case iterations of (2.19) are dense on the circle, they can be reduced by virtue of a smooth transformation $\varphi \to \bar{\varphi}$ to a circle shift

$$\bar{\varphi}_{n+1} = \bar{\varphi}_n + w \pmod 1 \,, \tag{2.20}$$

with an irrational w. A proper tool to distinguish these two regimes is the rotation number

$$\rho_\varphi = \lim_{N\to\infty} \frac{\varphi_N - \varphi_0}{N} \,, \tag{2.21}$$

which is rational in the periodic case and irrational in the quasiperiodic one (in the latter case $\rho = w$). Note that in the periodic case the Lyapunov exponent is negative, while for the quasiperiodic regime it is equal to zero (the latter fact trivially follows from (2.20)).

A quasiperiodically forced circle map is described by

$$\varphi_{n+1} = \varphi_n + g(2\pi\varphi_n, 2\pi\theta_n) \pmod 1 , \tag{2.22}$$

$$\theta_{n+1} = \theta_n + \omega \pmod 1 . \tag{2.23}$$

The phase space here is the two-dimensional torus $0 \leq \varphi, \theta < 1$. As a particular example we use $g(2\pi\varphi, 2\pi\theta) = a\sin(2\pi\varphi) + b\sin(2\pi\theta) + c$.

Consider first several simple dynamical regimes in the quasiperiodically forced circle map in relation to its dynamical counterpart in the autonomous case. The first is an attractor on the torus with a rational rotation number in φ-direction (2.21). This state can be considered as a quasiperiodic perturbation of a periodic orbit of the autonomous circle map. In the phase space it is represented by a curve wrapping the torus in θ-direction, or by a number of such curves (see Fig. 2.2a). This regime can be characterized as a quasiperiodic one with one incommensurate frequency ratio ω (in continuous time a corresponding regime is two-frequency quasiperiodic). A more complex state appears if an attractor in (2.22, 2.23) is a curve or a set of curves wrapping the phase space in a diagonal direction, i.e. in a skew direction with regard to θ and φ. Then the rotation number has in general the form $\rho_\varphi = \frac{m+n\omega}{l}$ with integers m, n, l. This state is also quasiperiodic with one incommensurate frequency ratio. Both described dynamical regimes are stable, i.e. they have a negative Lyapunov exponent.

In (2.22, 2.23) also a regime with irrational ρ_φ can be observed that via a smooth transformation can be reduced to a combination of two quasiperiodic rotations: in θ-direction this rotation is given by (2.23) and in φ-direction it is given by (2.20). In this regime the Lyapunov exponent is zero, here there are two incommensurate frequency ratios ω and ρ_φ, a corresponding continuous-time state would have three mutually incommensurate frequencies. This regime looks like in Fig. 2.2b.

Finally, in model (2.22, 2.23) a strange nonchaotic attractor can be observed. Here the Lyapunov exponent is negative, but the attractor does not look like a curve or a set of curves, see Fig. 2.2c.

2.2.3 *Skew shift*

Recently, Hunt and Ott [2001] considered the following modification of (2.22, 2.23)

$$\varphi_{n+1} = \varphi_n + \theta_n + g(2\pi\varphi_n, 2\pi\theta_n) \pmod 1 , \tag{2.24}$$

$$\theta_{n+1} = \theta_n + \omega \pmod 1 . \tag{2.25}$$

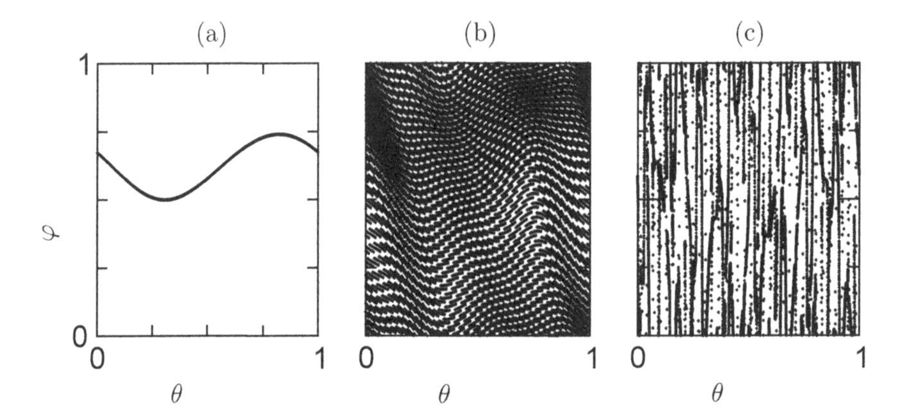

Fig. 2.2 Three different regimes in the quasiperiodically forced circle map. (a): A quasiperiodic state with two frequencies for $a = 0.05$, $b = 0.2$, $c = 0.03$. (b): A quasiperiodic state with three frequencies for $a = 0.05$, $b = 0.2$, $c = 0.15$. (c): An SNA for $a = 0.125$, $b = 3.1$, $c = 0$. In this regime the Lyapunov exponent is ≈ -0.2.

Now, due to the term θ_n in the r.h.s. of (2.24), one cannot consider the forcing as a small one. For vanishing g model (2.22, 2.23) reduces to a so-called skew shift, which is a popular model of weak irregularity in ergodic theory [Cornfeld et al. 1982]. Hunt and Ott [2001] have argued that for small nonvanishing g the attractor in (2.24, 2.25) is strange nonchaotic, see Section 5.1 below. The limitation of model (2.24, 2.25) is that due to it "skewness" (this means that for $\theta \gtrsim 0$ the phase φ shifts by a small value, while for $\theta \lesssim 2\pi$ it shifts by 2π) it can hardly be observed, contrary to (2.22, 2.23), in real continuous-time systems.

2.2.4 *Forced logistic map*

The logistic map is a paradigmatic example of chaos. An autonomous logistic map reads:

$$x_{n+1} = a - x_n^2 \ . \tag{2.26}$$

Sometimes one writes this map in another normalization, then the r.h.s. reads $rx_n(1 - x_n)$. This map is non-invertible and can demonstrate regular (typically periodic) and chaotic behavior. With a quasiperiodic forcing this map is usually written as

$$x_{n+1} = a - x_n^2 + \varepsilon \cos(2\pi\theta_n) \ , \tag{2.27}$$

$$\theta_{n+1} = \theta_n + \omega \ . \tag{2.28}$$

Various dynamical regimes that can be observed for different values of parameters a and ε, are depicted in Fig. 2.3.

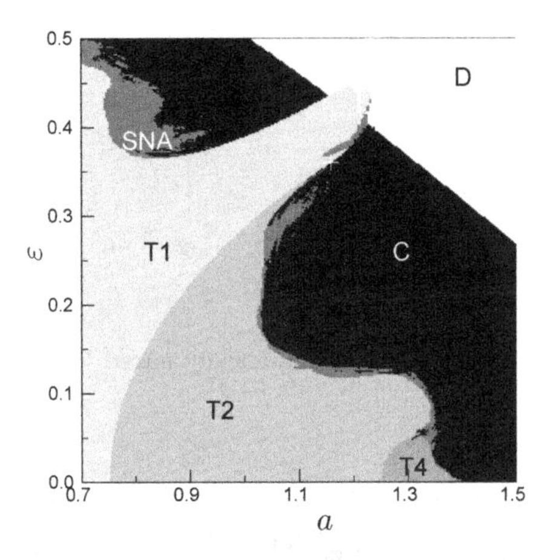

Fig. 2.3 Parameter plane for the forced logistic map (2.27, 2.28). D: for these parameter values a trajectory escapes to infinity. T1: a torus given by a single invariant curve (Fig. 2.4). T2: Doubled torus (a torus consisting of two invariant curves) (Fig. 2.5). T4: quadrupled torus (an invariant set consisting of four curves). C: Chaos (Fig. 2.6). SNA: strange nonchaotic attractor (Fig. 2.7).

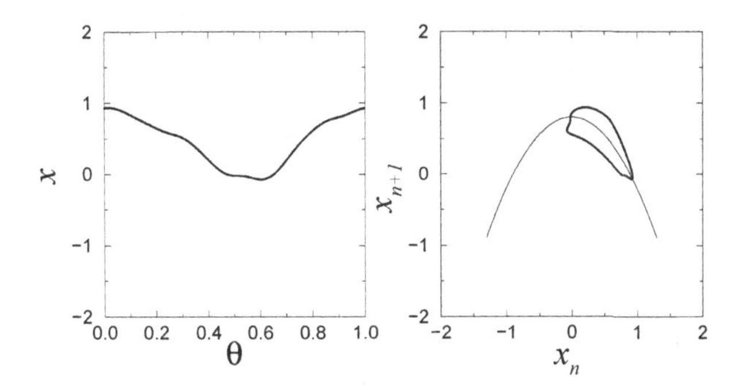

Fig. 2.4 A torus in the quasiperiodically forced logistic map (2.27, 2.28) for $a = 0.8, \varepsilon = 0.3$.

We describe qualitatively what is observed in model (2.27, 2.28) when changing parameters a, ε. For $\varepsilon = 0$ we have just an unforced logistic map, which demonstrates a transition to chaos via an infinite series of period-doubling bifurcations. For $0 < a < \frac{3}{4}$ this map has a stable fixed point, for $\frac{3}{4} < a < 1.25$ there is a stable period-2 orbit, etc. These stable

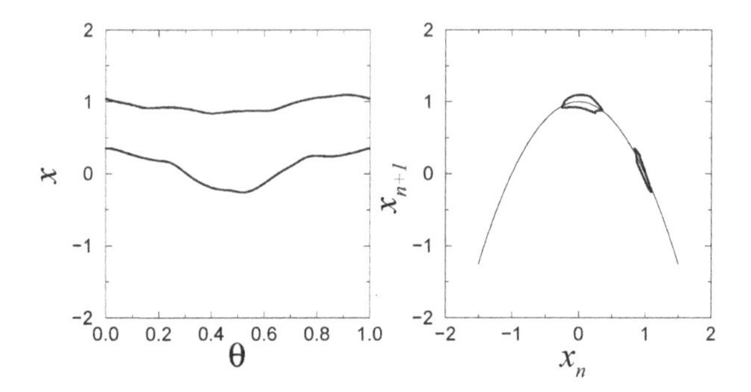

Fig. 2.5 A doubled torus in the quasiperiodically forced logistic map (2.27, 2.28) for $a = 1, \varepsilon = 0.1$.

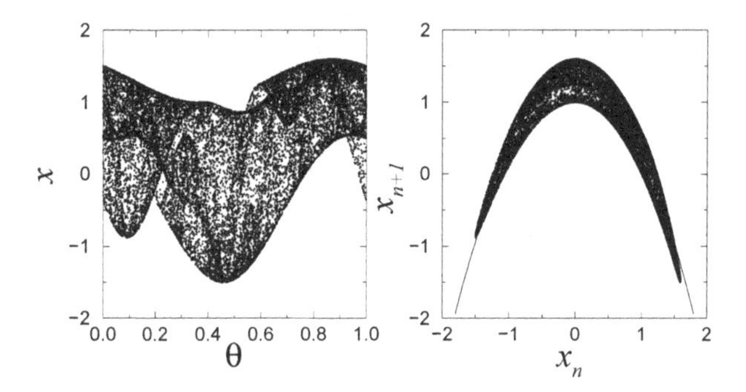

Fig. 2.6 Chaotic regime in the quasiperiodically forced logistic map (2.27, 2.28) for $a = 1.3, \varepsilon = 0.3$.

regimes survive small quasiperiodic forcing. For small ε a stable fixed point transforms to a torus, shown in Fig. 2.4. In the coordinates (x, θ) it is just a single-valued function, in the plane (x_n, x_{n+1}) this state is represented by a closed curve ("torus-1" or $T1$) around the fixed point. A stable period-2 orbit is also slightly disturbed for small quasiperiodic forcing, yielding a torus-2 (doubled torus or $T2$) Fig. 2.5). This torus in coordinates (x, θ) is represented by two curves and the trajectory at each iteration jumps from one curve to the other one; in coordinates (x_n, x_{n+1}) the attractor is represented by two closed invariant curves, a trajectory rotates on them and jumps from one to the other one. Similar regimes of torus-4 ($T4$), torus-8 ($T8$), etc. corresponding to period-4, period-8, etc. orbits can be also observed. A transition from torus-1 to torus-2 can be described as a

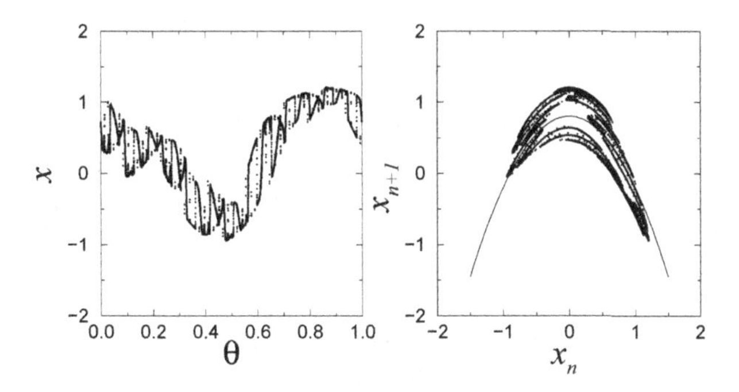

Fig. 2.7 An SNA in the quasiperiodically forced logistic map (2.27, 2.28) for $a = 0.8, \varepsilon = 0.4$.

torus-doubling, see Chapter 6 for details.

For large values of parameter a the logistic map exhibits chaos. This chaos is preserved under quasiperiodic forcing, moreover, for strong forcing (large values of ε) chaos can be observed already for smaller values of a. An example of a chaotic state is shown in Fig. 2.6.

Between the quasiperiodic motion on tori and chaotic behavior we find strange nonchaotic attractors like in Fig. 2.7. In both representations this attractor looks like a fractal, highly wrinkled set. That this regime is not chaotic is checked by calculating the Lyapunov exponent (Fig. 2.8): in the range $0.35 < \varepsilon < 0.45$ the exponent is negative. Generally, these different types of motion are typical for quasiperiodically forced systems demonstrating a transition to chaos: in the presence of a quasiperiodic forcing an SNA appears as an intermediate stage between order (torus) and chaos.

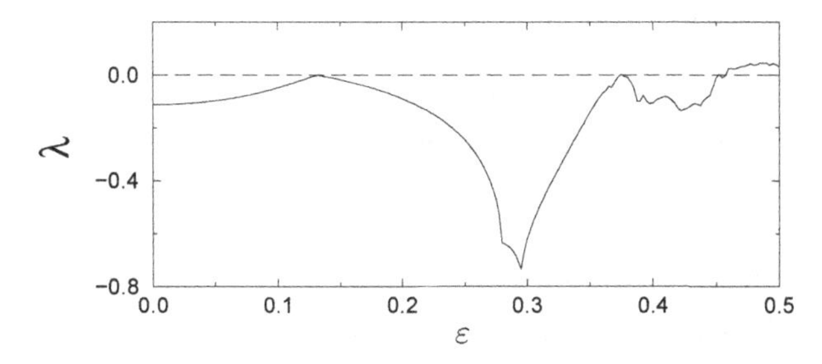

Fig. 2.8 Lyapunov exponent of the quasiperiodically forced logistic map (2.27, 2.28) depending on parameter ε for $a = 0.8$, calculated according to (2.10).

2.2.5 *Harper model*

There exists an interesting application of quasiperiodically forced one-dimensional maps to the theory of quasicrystals. The latter considers quasiperiodic lattices, in particular motions in quasiperiodic potentials. The simplest model of this type, a so-called Harper model, reads

$$\psi_{n+1} + \psi_{n-1} + V(2\pi\omega n)\psi_n = E\psi_n \,. \tag{2.29}$$

Here ψ_n is a wave function defined on discrete sites $-\infty < n < \infty$, $V(2\pi\omega n)$ is a 2π-periodic on-site potential that depends on n in a quasiperiodic manner provided ω is irrational, and E is the energy eigenvalue. One can interprete Eq. (2.29) as a stationary Schrödinger equation in a quasiperiodic potential. It is easy to check, that the Harper model is equivalent to a quasiperiodically forced one-dimensional map of type (2.8, 2.9)

$$x_{n+1} = E - \frac{1}{x_n} - V(2\pi\theta_n) \,, \tag{2.30}$$

$$\theta_{n+1} = \theta_n + \omega \pmod 1 \,, \tag{2.31}$$

where $x_n = \frac{\psi_n}{\psi_{n-1}}$. We will discuss properties of this map in Section 7.8.

2.3 Quasiperiodically forced high-dimensional maps

A great deal of work in nonlinear dynamics has been devoted to the investigation of the transition to chaos in higher-dimensional maps, in particular in two-dimensional maps. This is due to the fact that two-dimensional invertible maps can be considered as an equivalent to a Poincaré section of a three-dimensional flow. Thus, all phenomena which are observable in two-dimensional invertible maps including the transition to chaos can be found in flows too. For this reason we also focus on the study of two-dimensional quasiperiodically forced maps to draw conclusions about the appearance of strange nonchaotic attractors and the subsequent transition to chaos in flows.

The simplest case of a two-dimensional map with quasiperiodic forcing is the Hénon map:

$$\begin{aligned} x_{n+1} &= 1 - bx_n^2 + y_n + \varepsilon\cos(2\pi\theta_n) \,, \\ y_{n+1} &= cx_n \,, \\ \theta_{n+1} &= \theta_n + \omega \pmod 1 \,, \end{aligned} \tag{2.32}$$

where b denotes the nonlinearity and ε measures the strength of the forcing. The autonomous Hénon map is known to show a period doubling route into chaos. With nonzero forcing, we observe several types of bifurcations, such as torus doublings [Sosnovtseva et al. 1996], boundary crisis and basin boundary metamorphosis [Osinga and Feudel 2000], which will be discussed in Chapter 6. Here we only demonstrate the various types of attractors in the Hénon map in different projections (Fig. 2.9). These attractors appear when the nonlinearity parameter b or the strength of the forcing ε are varied.

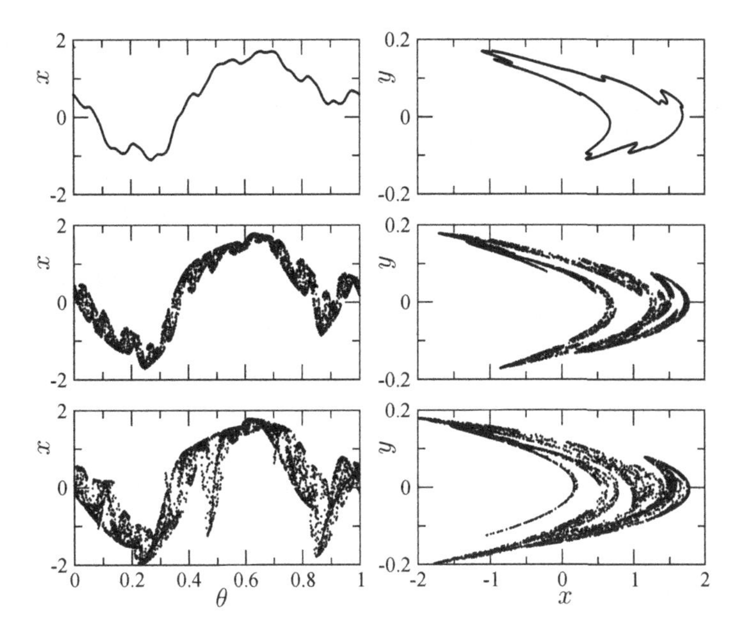

Fig. 2.9 Attractors in the quasiperiodically forced Hénon map (2.32) with $c = 0.1$, $\varepsilon = 0.7$ projected to the (x, y) space (right column) and to the (θ, x) space (left column): quasiperiodic motion on a torus (top row) for $b = 0.6$, strange nonchaotic attractor (middle row) for $b = 0.7$, and chaotic attractor (bottom row) for $b = 0.77$.

2.4 Quasiperiodically forced continuous-time systems

In this section we present several examples of realistic continuous-time physical systems described by equations of type (2.1-2.3).

2.4.1 *Forced overdamped pendulum*

The simplest case of system (2.1-2.3) is that with a one-dimensional state vector $\mathbf{x} = x$. In such a system chaos cannot occur – here the same argument as used in Sec. 2.2 above, concerning the non-existence of chaos in monotonous maps, is valid, because a one-dimensional differential equation reduces to a monotonous one-dimensional map. As an example we consider the following equation:

$$\dot{x} = -\sin x + b_0 + b_1 \sin(\omega_1 t) + b_2 \sin(\omega_2 t) \ . \tag{2.33}$$

It describes an overdamped rotator (pendulum) subject to two forces with frequencies ω_1 and ω_2 (and also a constant force b_0). There is also another interpretation of Eq. (2.33) – it describes a resistively shunted Josephson junction subject to an external current consisting of two periodic components and a constant current proportional to b_0. In this latter interpretation the derivative $\dot{x}$ is proportional to the voltage across the junction.

Because the variable x can be considered as 2π-periodic, Eq. (2.33) can be reduced to a forced circle map like (2.22, 2.23). Correspondingly, three types of dynamical regimes depicted in Fig. 2.2 can be observed here: quasiperiodic motions with two and three incommensurate frequencies as well as strange nonchaotic behavior. In Fig. 2.10 we show a part of the parameter space, namely the region of parameters b_1, b_2 where only two-frequency motion (white dots) and SNA (black dots) occur in (2.33) for $b_0 = 0$ and $\omega_1 = 1$, $\omega_2 = \frac{\sqrt{5}-1}{2}$. Time series of these regimes are presented in Fig. 2.11.

The SNA regime in model (2.33) has one remarkable property that allows one to distinguish it easily from a two-frequency torus. One does not need to construct the phase portrait like Fig. 2.2, but just to look on the average characteristics of motion. For Josephson junctions such characteristics are the dc components of the external current b_0 and the voltage $\langle \dot{x} \rangle$. If for $b_0 = 0$ there exists a stable torus then the variable x is bounded and $\langle \dot{x} \rangle = 0$. For small $b_0 \neq 0$ the torus still exists – it can disappear only through a bifurcation with a creation either of a quasiperiodic state with three frequencies (like in the middle panel of Fig. 2.2) or of an SNA. But unless this bifurcation occurs, the variable x remains bounded (see Fig. 2.11) and $\langle \dot{x} \rangle = 0$. This means that zero voltage $\langle \dot{x} \rangle$ corresponds to nonzero current b_0, i.e. one observes a supercurrent and the state of the Josephson junction is superconducting in some range of currents b_0. Another situation occurs if the state at $b_0 = 0$ is an SNA. Here the variable x changes

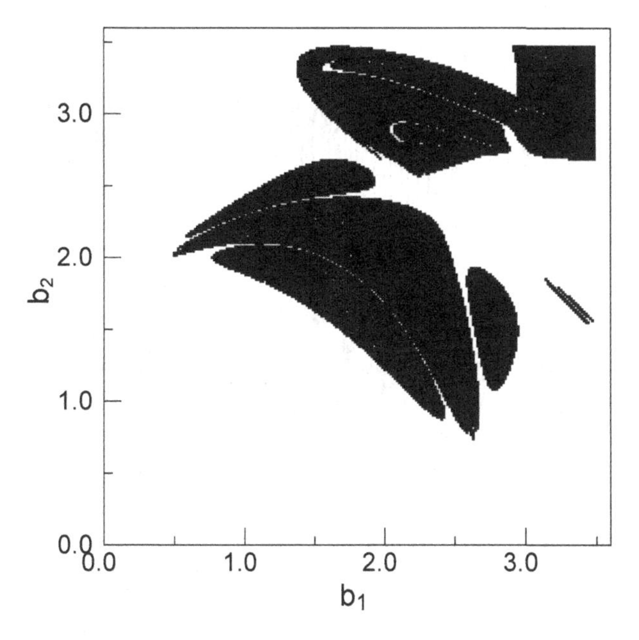

Fig. 2.10 The region in the plane of parameters of system (2.33) at which an SNA is observed ($b_0 = 0, \omega_1 = 1.0, \omega_2 = \frac{\sqrt{5}-1}{2}$). Regions corresponding to SNA are marked black, white regions correspond to a quasiperiodic motion with two incommensurate frequencies.

unboundedly (see Fig. 2.11), although due to the symmetry $x \to -x$ the average voltage vanishes $\langle \dot{x} \rangle = 0$. However, already a small current $b_0 \neq 0$ breaks the symmetry and x drifts in one direction, i.e. means that $\langle \dot{x} \rangle \neq 0$ and the voltage is non-zero. Thus, the state of the Josephson junction with an SNA is resistive: a non-zero current corresponds to a non-zero voltage. We show these two distinct current-voltage characteristics for model (2.33) in Fig. 2.12.

2.4.2 *Forced Duffing oscillator*

If in a continuous-time model (2.1-2.3) the dimension of $\mathbf{x}$ is two or larger, then chaotic states are possible in such a system. The situation here is similar to that in the quasiperiodically forced logistic map (see Section 2.2.4): as a parameter is changed, generally a transition from order to chaos is observed. In a quasiperiodically forced system "order" means a quasiperiodic regime with a torus as an attractor; chaos is characterized by a positive maximal Lyapunov exponent. Usually, near the transition to chaos a

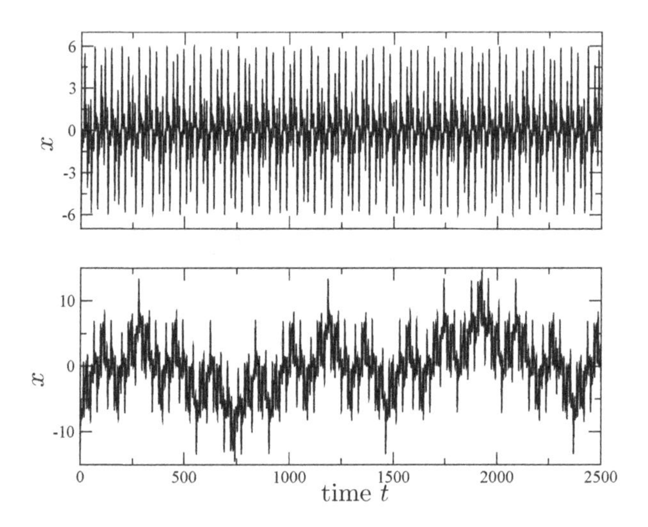

Fig. 2.11 Time series $x(t)$ of (2.33) corresponding to a quasiperiodic motion (top panel, $b_1 = b_2 = 1.5$, $b_0 = 0$) and to SNA (bottom panel, $b_1 = b_2 = 2$, $b_0 = 0$).

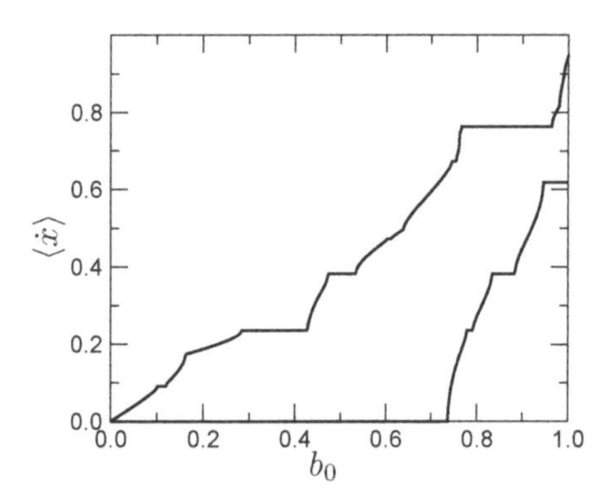

Fig. 2.12 Current-voltage characteristics of the quasiperiodically forced Josephson junction (2.33). Lower curve: torus at $b_0 = 0$, $b_1 = b_2 = 0.5$, here the state at zero current is superconducting. Upper curve: SNA at $b_0 = 0$, $b_1 = b_2 = 2$, this state is resistive. The horizontal steps with $\langle \dot{x} \rangle \neq 0$ correspond to tori with rotation numbers different from zero.

strange nonchaotic attractor is observed, like in the forced logistic map.

We illustrate this with a quasiperiodically forced Duffing oscillator considered by Heagy and Ditto [1991]. This model reads

$$\frac{d^2x}{dt^2} + 0.1\frac{dx}{dt} - x + x^3 = ax(\cos\phi_1 + 0.3\cos\phi_2)\,, \tag{2.34}$$

$$\frac{d\phi_1}{dt} = \frac{\sqrt{5}+1}{2}\,, \tag{2.35}$$

$$\frac{d\phi_2}{dt} = 1\,. \tag{2.36}$$

As described in Section 2.1, a visual representation of the attractors in this model can be achieved by using a stroboscopic map with time interval $T = \frac{4\pi}{\sqrt{5}+1}$, i.e. the period of the phase ϕ_1. The corresponding mapping $(x_n, \dot{x}_n, \theta_n) \to (x_{n+1}, \dot{x}_{n+1}, \theta_{n+1})$ is three dimensional, but it is convenient to present its projection on the plane (x_n, θ_n). This projection of attractors is shown in Fig. 2.13.

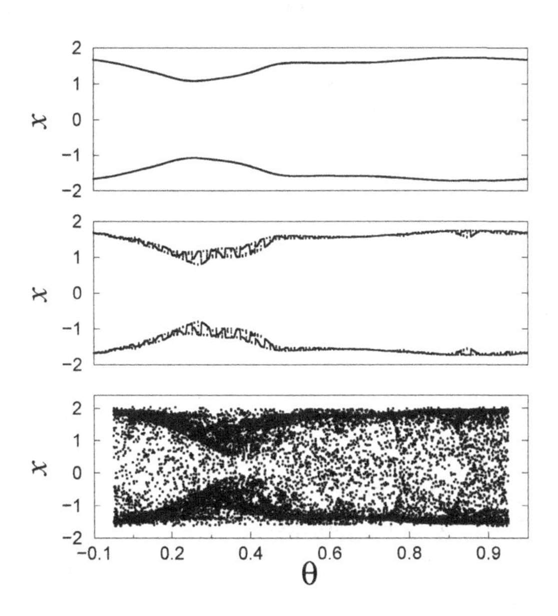

Fig. 2.13 Attractors in model (2.34-2.36). Top panel: a torus for $a = 0.38$. Middle panel: an SNA for $a = 0.396$. Bottom panel: a chaotic attractor for $a = 0.41$.

2.5 Experiments

Attempts to observe strange nonchaotic attractors have been performed in several experiments. Typically, one takes a nonlinear dissipative system and applies a quasiperiodic force. Then one has to verify that the observed attractor is strange and nonchaotic. The first task is relatively simple, as one can construct the stroboscopic map from the observed data and prove that it is more complex than a torus. To prove nonchaoticity one has, however, to prove that the largest Lyapunov exponent is negative, what can be hardly done experimentally. Therefore one relies on other characteristics of strange nonchaotic dynamics like spectra, dimensions, etc. We will see in the next chapters how these characteristics work in numerical studies; it will be clear how difficult it is to apply them to real observations.

The first, to the best of our knowledge, experiment with a strange nonchaotic attractor has been performed by Ditto et al. [1990] with a forced magnetic ribbon. Zhou et al. [1992] studied experimentally an SNA in a quasiperiodically forced electronic circuit with a multistable potential, modeling a SQUID device. Zeyer et al. [1995] forced quasiperiodically a Belousov-Zhabotinsky reaction, using a superposition of signals from two other chemical reaction cells operating periodically as a force. A simple electronic circuit with a two-frequency forcing has been studied by Yang and Bilimgut [1997].

In several experimental works one attempted to follow theoretical predictions. Yu et al. [1999b] tried to realize a transition to SNA through a modulated pitchfork bifurcation. Venkatesan et al. [1999] considered a quasiperiodically forced nonlinear electronic circuit and compared experimentally observed phase portraits and power spectra with numerical data. In papers [Bezruchko et al. 2000; Seleznev and Zakharevich 2005] a method of periodic (rational) approximations to a quasiperiodic force (see Chapter 3) has been used. Instead of applying a quasiperiodic signal, in these experiments a nonlinear electronic circuit was forced with complex periodic signals. The experimental findings have been compared with the corresponding numerical simulations for periodic approximations, resulting in a very reliable evidence for a strange nonchaotic behavior.

2.6 Bibliographic notes

As already mentioned, investigations of SNAs have been pioneered by Grebogi, Ott, Pelikan, and Yorke [1984]. Simultaneously Kaneko [1984a] studied a quasiperiodically forced circle map and observed complex attractors with negative Lyapunov exponents. Furthermore, he studied a quasiperiodically forced logistic map and observed the fractalization of the torus [Kaneko 1984b], later Nishikawa and Kaneko [1996] performed a more detailed analysis of this problem[1]. For other studies of a circle map see [Ding et al. 1989b; Kapitaniak et al. 1990; Feudel et al. 1995a; Glendinning 1998; Sturman 1999a; Glendinning et al. 2000; Osinga et al. 2001; Stark et al. 2002]. The logistic map has been considered by Witt, Feudel, and Pikovsky [1997]; Prasad, Mehra, and Ramaswamy [1997, 1998]; Negi, Prasad, and Ramaswamy [2000]; Khovanov, Khovanova, Anishchenko, and McClintock [2000a]; Kim, Lim, and Ott [2003b]; Lim and Kim [2005]; Kim and Lim [2005]. Venkatesan and Lakshmanan [2001] studied a cubic one-dimensional map. Hunt and Ott [2001]; Kim et al. [2003a] considered an SNA of skew shift type. Chandre et al. [1999] studied a three-frequency renormalization transformation for a Hamiltonian system with three degrees of freedom and have found an SNA there.

A relevance of SNAs for the eigenvalue problem in quasiperiodic potentials have been first mentioned by Bondeson, Ott, and Antonsen [1985], who showed that a Schrödinger equation with a quasiperiodic potential is equivalent to a forced overdamped pendulum, possessing an SNA. Further studies have been mainly focused on the Harper [Ketoja and Satija 1995, 1997b; Prasad et al. 1999; Negi and Ramaswamy 2001a; Datta et al. 2003; Mestel and Osbaldestin 2004; Datta et al. 2005] and Frenkel-Kontorova [Ketoja and Satija 1997a] models. We mention here that complex dynamics has been also observed in many quasiperiodically forced quantum systems [Shepelyansky 1983; Samuelides et al. 1986; Pomeau et al. 1986; Sutherland 1986; Badii and Meier 1987; Blekher et al. 1992; Luck et al. 1988; Geisel 1990; Crisanti et al. 1994; Feudel et al. 1995b; Pikovsky et al. 1996], although a relation to SNAs is not clear for these non-dissipative models.

Maps in two and higher dimensions have been considered by Sosnovtseva et al. [1996]; Anishchenko et al. [1996]; Sosnovtseva et al. [1998]; Vadivasova et al. [1999]; Jalnine and Osbaldestin [2005]. Different continuous-time

[1] As already mentioned, SNAs are observed in quasiperiodically forced systems, so we omit "quasiperiodically forced" in the descriptions below. Some cases of driving more complex than a quasiperiodic one have been discussed by Negi and Ramaswamy [2001b].

models have been studied in [Romeiras and Ott 1987] (pendulum), [Kapitaniak et al. 1990; Pokorny et al. 1996] (Van der Pol oscillator), [Neumann and Pikovsky 2002; Kuznetsov and Neumann 2003] (Josephson junction model), [Yagasaki 1999; Khovanov et al. 2000b; Venkatesan et al. 2000] (Duffing oscillator), [Prasad et al. 2003] (an excitable system); see also [Venkatesan and Lakshmanan 1997, 1998; Yu et al. 1999a]. Stagliano et al. [1996] studied two coupled Duffing systems with forces having different frequencies. Ramaswamy [1997] considered two coupled multistable oscillators similar to that in experiments [Zhou et al. 1992]. Kapitaniak [1993] discussed a general continuous-time dynamics in the presence of a quasiperiodic forcing consisting of periodic pieces.

Effects of noise were discussed by Khovanov et al. [2000b,a]; Kuznetsov [2005]. A situation, where a strange nonchaotic attractor is weakly attracting (is a Milnor attractor) has been studied by Yang [2001]. A review on strange nonchaotic attractors was given by Prasad, Negi, and Ramaswamy [2001].

Chapter 3

Rational approximations

The method of rational approximations is based on the fact that all irrational numbers can be approximated by appropriate rationals. Therefore it is quite natural to use this property to describe quasiperiodically forced systems in terms of rational approximations. This approach is well-known from studies of resonance phenomena in Hamiltonian (KAM theory)[Greene 1979; MacKay 1983] and dissipative systems (transition to chaos through quasiperiodicity) [Feigenbaum et al. 1982a; Rand et al. 1982].

The method of rational approximations is quite useful from different perspectives. Firstly, from a physical viewpoint a distinction between rational and irrational numbers is rather subtle, as in an experiment (and even in numerics) one can hardly distinguish them. Thus it is quite helpful to have a description that deals with rational numbers only. In particular, following rational approximations may be a proper way to study quasiperiodically forced systems experimentally. Secondly, as we will see below, on the basis of rational approximations one can derive criteria which can be used to detect SNAs and to distinguish them from chaotic attractors. Looking at the phase portraits of SNAs in various examples of quasiperiodically forced systems it becomes obvious that it is rather difficult to sort out whether a particular attractor is smooth, but very wrinkled, strange nonchaotic, or strange chaotic. Thus conditions are needed which enable us to make a clear distinction between the different kinds of attractors. The investigation of rational approximations of the attractors is a way to formulate such conditions as it will be discussed in this chapter. Last but not least, rational approximations are a very useful tool to study the transitions to SNA when a system's parameter is varied. Examples how to use this method for the purpose of identifying the mechanisms of the emergence of SNAs are presented in Chapter 6.

As we have already discussed in Chapters 1 and 2, a quasiperiodic forcing can be realized in two different forms when studying flows and maps. In flows which are described by ordinary differential equations quasiperiodic forcing is usually achieved by applying two external periodic signals with amplitudes a_1 and a_2 and frequencies ω_1 and ω_2. In general, one can add a term $a_1 \sin(\omega_1 t) + a_2 \sin(\omega_2 t)$. The two frequencies ω_1 and ω_2 are incommensurate, i.e. they do not fulfill a resonance condition. Thus it holds $\omega_1/\omega_2 \neq p/q$ with p and q integers. In maps the quasiperiodic forcing is realized by a rigid rotation of a phase variable θ with a particular frequency w being an irrational number. The simplest way of realization is adding a map like $\theta_{n+1} = \theta_n + w$. Looking at Poincaré sections in flows both representations of quasiperiodic forcing become equivalent with $w = \omega_1/\omega_2$. Having this equivalence in mind (see also Section 2.1), we can focus our discussion of the method of rational approximations to the description of maps.

3.1 Properties of rational approximations of irrationals

Rational approximations w_k of an irrational number w are obtained by using its continued fraction representation (for details see any book on number theory, in particular [Khinchin 1949; Rocket and Szüsz 1992]). Each irrational $0 < w < 1$ can be represented as an infinite continued fraction

$$w = \cfrac{1}{a_1 + \cfrac{1}{a_2 + \cdots}} = [a_1, a_2, ...] \tag{3.1}$$

with integers a_k. Rational approximations for the irrational are obtained if one takes the so-called k-th convergents of the fraction:

$$w_k = \frac{p_k}{q_k} = [a_1, a_2, ..., a_k] = \cfrac{1}{a_1 + \cfrac{1}{a_2 + \cfrac{1}{\cdots + \cfrac{1}{a_k}}}}, \tag{3.2}$$

where p_k and q_k satisfy the recursion relations

$$\begin{aligned} &p_k = a_k p_{k-1} + p_{k-2}, \quad q_k = a_k q_{k-1} + q_{k-2}, \\ &p_0 = 0, \; q_0 = p_1 = 1, \; q_1 = a_1. \end{aligned} \tag{3.3}$$

These rational numbers w_k converge to the irrational number w in the limit $k \to \infty$: $w = \lim_{k\to\infty} w_k$. The higher k, the better is the approximation of the irrational number. The convergents w_k lie alternatively above and below w: even convergents form an increasing sequence approaching w from below, while odd convergents form a decreasing sequence approaching w from above. The rate of convergence is estimated as follows:

$$\frac{1}{(a_{k+1}+2)q_k^2} < |w - w_k| < \frac{1}{a_{k+1}q_k^2} \ . \tag{3.4}$$

An important question is: how good a particular irrational number is approximated by the series of w_k. From the properties above it is clear, that if the numbers a_k are large, the sequence of q_k grows rapidly and the irrational w is extremely close to its approximations w_k. Such irrationals are from the practical (say, experimental or numerical) viewpoint like rationals. By contrast, if the numbers a_k are small, the irrational number is rather different from its rational approximations and such a number can be viewed as a "good" irrational. Mathematically, such good irrationals are called Diophantine numbers.

One of the most prominent irrational numbers is the inverse of the golden mean $\omega = (\sqrt{5}-1)/2$. Its continued fraction representation can be written as $\omega = [1, 1, 1, ...]$ with

$$\omega_k = \frac{p_k}{q_k} = \frac{F_{k-1}}{F_k} \ , \tag{3.5}$$

where $F_k = 1, 1, 2, 3, 5, 8, 13, 21, \ldots$ are the Fibonacci numbers satisfying, according to (3.3), the recursion relation $F_k = F_{k-1} + F_{k-2}$. According to the discussion above, the inverse of the golden mean is the "best" irrational (often called the most irrational number), as here the numbers a_k are minimal, and thus ω is maximally different from its rational approximations. Due to this property it plays a special role in the theory of dynamical systems: the last KAM torus to disappear in Hamiltonian systems corresponds to a quasiperiodic motion with this frequency [Lichtenberg and Lieberman 1992]. One also introduces the silver mean with $w = [2, 2, 2, ...]$, and other irrationals represented by periodic continued fractions. All such numbers can be expressed via square roots of integers.

3.2 Rational approximations to quasiperiodic forcing

We now discuss, how a quasiperiodic forcing can be approximated with periodic ones. The equation for the forcing phase θ_n which for an irrational w reads

$$\theta_{n+1} = \theta_n + w \pmod 1 \tag{3.6}$$

should be replaced by

$$\theta_{n+1} = \theta_n + w_k \pmod 1 \tag{3.7}$$

with $w_k = p_k/q_k$. The crucial difference between (3.6) and (3.7) lies in the following. The trajectory of (3.6) is infinite and the points θ_n are dense on the interval $0 \leq \theta < 1$ because for irrational w all the points $w,\ 2w,\ 3w, \ldots$ taken modulo 1 are different. By contrast, a trajectory of (3.7) is finite and consists of q_k points, because

$$\theta + q_k w_k = \theta + p_k = \theta \pmod 1 .$$

If we want to approximate the whole interval $0 \leq \theta < 1$ of values of the forcing phase like in (3.6) with (3.7), we have to consider a family of systems (3.7). It is clear that this family can be parameterized by the value θ_0 – the initial phase, hereafter we call it *phase shift*. Indeed, choosing different phase shifts θ_0 we obtain according to (3.7) different trajectories of length q_k. It is clear that it is sufficient to choose $0 \leq \theta_0 < 1/q_k$, then the corresponding periodic trajectories will fill the interval $0 \leq \theta < 1$.

As a result, *one* quasiperiodically forced system is approximated by a *family* of periodically forced systems where the irrational frequency w is replaced by its rational approximation w_k and the phase shift $\theta_0 \in [0, 1/q_k)$ is an additional parameter. If we analyze these systems for every k, then we expect that the properties of the original quasiperiodically forced system can be obtained by taking the limit $k \to \infty$. This way we derive powerful test criteria to check for the existence of SNA as well as analyze the emergence of SNA when a system's parameter is varied.

3.3 Checking strangeness of SNA through rational approximations

3.3.1 *Rational approximations to a smooth attractor*

We start a demonstration of the rational approximations approach to quasiperiodically forced systems with the simplest case, when the attractor is a smooth torus. Here and below we consider only the case of forcing with the inverse of the golden mean, where the rational approximations are given by the ratios of Fibonacci numbers (3.5). As a model we take system (2.17, 2.18) which for convenience is rewritten here

$$x_{n+1} = 2a\tanh(x_n)(b + \cos(2\pi\theta_n)) \ , \tag{3.8}$$

$$\theta_{n+1} = \theta_n + \omega \ . \tag{3.9}$$

We restrict ourselves for the moment to the case $b > 1$ which corresponds to a stable smooth torus (in fact, due to the symmetry $x \to -x$ there are two symmetric stable tori, for clarity of presentation we consider positive x only).

To find the k-th approximation to the attractor in this model, one proceeds as follows:
(i) the forcing term (3.9) is replaced with the periodic forcing (3.7);
(ii) for each initial phase θ_0 an attractor in the periodically forced mapping (3.8, 3.7) is determined;
(iii) the obtained family of these attractors is plotted for all $\theta_0 \in [0, 1/q_k)$, the graph obtained is the k-th approximation to the attractor in (3.8, 3.9).

For $\omega_1 = 1$ the period of the forcing is 1 and the attractor is the fixed point satisfying $x = 2a\tanh(x)(b + \cos(2\pi\theta_0))$. This equation gives a curve $x = x(\theta_0)$ which is the first approximation to the attractor (see Fig. 3.1). For $\omega_2 = 1/2$ the forcing has period two, now an attractor in (3.8, 3.7) consists of two points for each initial phase. Varying θ_0 in the interval $[0, 1/2)$ we obtain the second approximation (Fig. 3.1), etc. All these approximations converge to a curve – the smooth torus – that can be obtained via direct iterations of (3.8, 3.9).

3.3.2 *Rational approximations to an SNA: An example*

Now we apply the same approach to an SNA. We take the classical GOPY-map (2.11,2.12) (corresponding to $b = 0$ in (3.8)); for convenience we rewrite

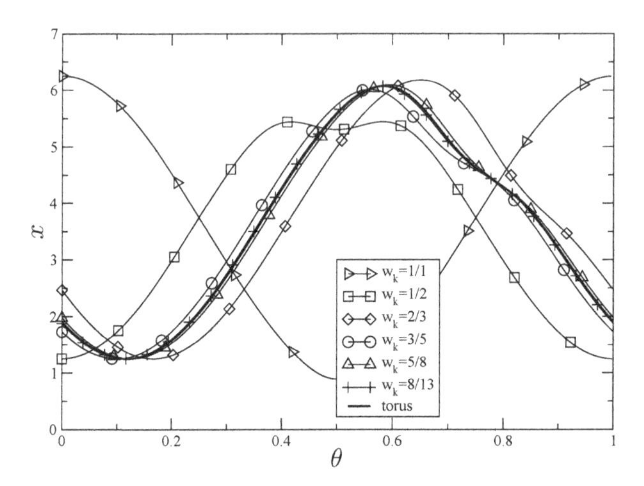

Fig. 3.1 A sequence of rational approximations to the smooth torus in system (3.8, 3.9).

it here for the case of periodic driving

$$x_{n+1} = 2a \tanh(x_n) \cos(2\pi\theta_n) , \tag{3.10}$$

$$\theta_{n+1} = \theta_n + \omega_k . \tag{3.11}$$

Again, the lowest approximation $\omega_1 = 1$ where the driving force is a constant is the simplest one and can be treated analytically. It reduces to a one-dimensional mapping

$$x_{n+1} = 2a \tanh(x_n) \cos(2\pi\theta_0) \tag{3.12}$$

depending on θ_0 as on a parameter. In fact, we have to find the bifurcation diagram – the attractors in (3.12) for different values of the parameter. For $0 \leq \theta_0 \leq 1/4$ and $3/4 \leq \theta_0 \leq 1$ the factor $\lambda = 2a\cos(2\pi\theta_0)$ is positive (we assume $a > 0$). In this region if $\lambda < 1$, mapping (3.12) has one stable fixed point $x = 0$; if $\lambda > 1$, there are two symmetric stable fixed points (solutions of $x = \lambda \tanh x$) and an unstable fixed point $x = 0$. In the region $1/4 \leq \theta_0 \leq 3/4$ the factor λ is negative. Here mapping (3.12) has one stable fixed point $x = 0$ if $|\lambda| < 1$ and a stable period-2 cycle otherwise. This bifurcation diagram is shown in Fig. 3.2.

We now proceed to higher approximations. For $\omega_2 = 1/2$ the forcing in (3.10, 3.11) has period two. In order to look for attractors it is convenient to reduce this system to an autonomous mapping of type (3.12), taking

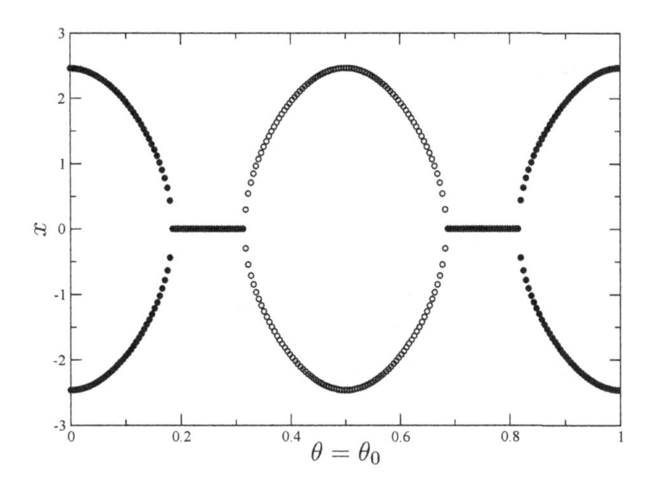

Fig. 3.2 Bifurcation diagram for map (3.12) with $a = 1.25$. Only stable attractors are shown: stable fixed points with filled circles, stable period-2 cycle with open circles.

each second time step. Simple calculations yield

$$x_{n+2} = -2a \tanh[2a \tanh(x_n) \cos(2\pi\theta_0)] \cos(2\pi\theta_0) \ . \tag{3.13}$$

This autonomous mapping can be treated similar to (3.12), it has stable fixed points and stable period-2 cycles. A stable fixed point of (3.13) is a period-2 orbit of (3.10, 3.11) and a stable period-2 orbit of (3.13) is a period-4 orbit of (3.10, 3.11). All these points give the 2nd approximation to the attractor in (3.8, 3.9), see Fig. 3.3.

¿From the discussion above it is clear that the higher approximations can be treated in a similar way. The 4th approximation with $\omega_4 = \frac{3}{5}$ leads to Fig. 3.4. Here the period of the driving force is 5 and one should classify attractors according to their behavior in the 5th iteration of (3.10, 3.11): there again we observe fixed points and period-2 cycles, which in normal time correspond to period-5 and period-10 orbits. In fact, the bifurcation structure of the rational approximation can be best seen if we restrict ourselves to the 5th iteration of (3.10, 3.11) and plot only a part of Fig. 3.4 corresponding to the basic interval of the phase shift θ_0, i.e. to the interval $[0, 1/q_k)$. This diagram is shown in Fig. 3.5.

Increasing the order of approximation we obtain new structures, in Figs. 3.6, 3.7 we show the 7th approximation with $\omega_7 = 13/21$. One can see that Fig. 3.6 represents the main features of the attractor already rather good (cf. Fig. 2.1). Fig. 3.7 shows that in this approximation the same bifurcations as in the lower-order approximations are observed.

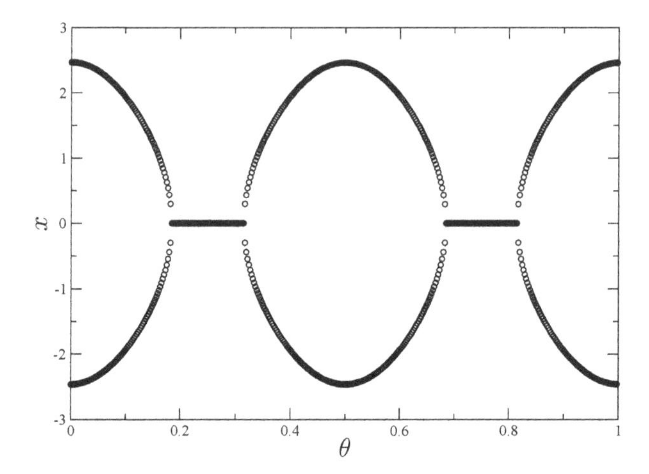

Fig. 3.3 The same as Fig. 3.2 but for the 2nd approximation $\omega_2 = 1/2$.

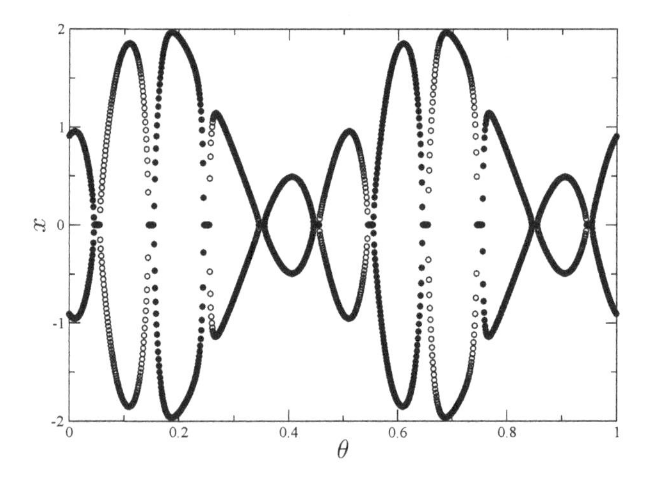

Fig. 3.4 The same as Fig. 3.2 but for the 4th approximation $\omega_4 = 3/5$.

3.3.3 *Rational approximations to an SNA: General consideration*

Above we have focused on a particular example of rational approximations. Hoping that the main ideas have been clarified, we now formulate them in

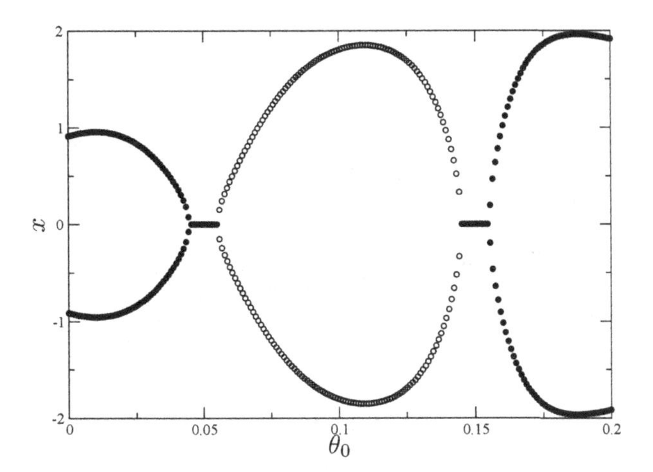

Fig. 3.5 The same as Fig. 3.4 but restricted to the basic interval of definition of the phase shift θ_0: $0 \leq \theta_0 < 1/5$.

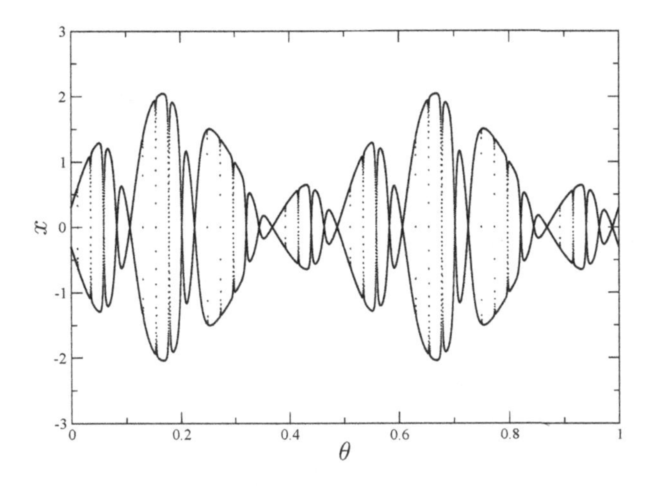

Fig. 3.6 The 7th approximation $\omega_7 = 13/21$.

a general form. The original system reads

$$\mathbf{x}_{n+1} = \mathbf{f}(\mathbf{x}_n, \theta_n) , \tag{3.14}$$

$$\theta_{n+1} = \theta_n + w \pmod 1 , \tag{3.15}$$

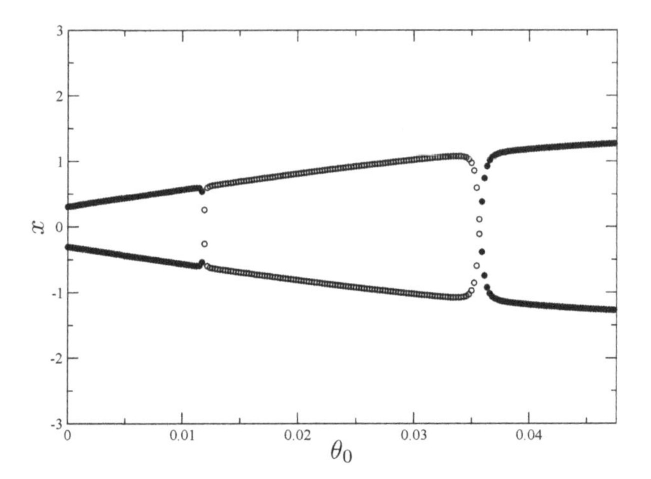

Fig. 3.7 The same as Fig. 3.6 but restricted to the basic interval of definition of the initial phase θ_0. Note that although possible attractors here are the same as in Fig. 3.5, the region of stability of the fixed point $x = 0$ is not seen due to its extreme smallness.

and its rational approximations are

$$\mathbf{x}_{n+1} = \mathbf{f}(\mathbf{x}_n, \theta_n) \ , \tag{3.16}$$

$$\theta_{n+1} = \theta_n + w_k \pmod 1 \ . \tag{3.17}$$

Suppose we look at the family of systems periodically forced with w_k (3.16, 3.17), parameterized by the phase shift θ_0. Each of these systems (3.16, 3.17) is a periodically (with period q_k) forced nonlinear map. Or, if we consider only each q_k-th point (some kind of stroboscopic observation), then the system is governed by an autonomous nonlinear map. This map is smooth and may exhibit different long-term behaviors such as fixed points, periodic or quasiperiodic orbits as well as strange chaotic attractors (the last possibility, however, might be excluded for particular nonlinear functions $\mathbf{f}$). The attracting sets in the periodically forced systems (3.16, 3.17) depend on the system parameters, and, what is even more important, on the *phase shift* θ_0 (the initial value of the phase in the driving (3.17)). It is clear, that it is sufficient to change θ_0 in the interval $[0, 1/q_k)$ in order to get all possible attracting sets in system (3.16, 3.17), because in the case of a q_k periodic orbit the set of all θ values fills the whole interval $[0, 1)$. Changing θ_0 continuously in this basic interval $[0, 1/q_k)$ and drawing the attracting set in the $(\theta, \mathbf{x})$ space for each of the chosen initial phases θ_0, we obtain the k–th approximation of the attractor in the quasiperiodically forced system

(3.14, 3.15) as the union of all occurring attracting sets. Investigating these approximations and taking the limit $k \to \infty$, we can classify the properties of the limiting attractor.

Since in Eqs. (3.14) and (3.16) $\mathbf{f}$ is supposed to be a nonlinear function, different θ_0 may lead to different attractors. Thus we can consider the phase shift θ_0 as a bifurcation parameter. We argue that the analysis of these bifurcation diagrams yields a criterion to distinguish between strange and nonstrange (smooth) attractors: Assuming that chaos is absent, *i.e.* no positive Lyapunov exponents occur, it is asserted that a sufficient condition for the appearance of SNA is the presence of bifurcations depending on the phase shift θ_0 in the dynamics of the rational approximations of the quasiperiodically forced system. Additionally, these bifurcations have to persist with increasing order of the approximations. In other words, the attractor in the quasiperiodically forced system (3.14, 3.15) is *strange*, if we obtain bifurcations depending on θ_0 at smaller and smaller scales with better and better approximation of the irrational w. This means that each of the periodically forced systems (3.16, 3.17) shows a more complex bifurcation diagram as k increases. If there are *no bifurcations* for a sufficiently large k, the attractor is *smooth*. In practice, smooth attractors usually may exhibit bifurcations depending on the phase shift θ_0 for a low order of the rational approximations, but these bifurcations disappear as the order of the approximation increases, *i.e.* in the limit $k \to \infty$.

The same approach can be used to check the smoothness properties of attractors in quasiperiodically forced differential equations. In this case one replaces the quasiperiodic forcing of type $a_1 \sin(\omega_1 t) + a_2 \sin(\omega_2 t)$ with a periodic one $a_1 \sin(\omega_1 t) + a_2 \sin(\frac{p}{q}\omega_1 t + \phi_0)$. Here ϕ_0 describes the phase shift between the two components and plays the role of the bifurcation parameter θ_0. One has to search for bifurcations depending on ϕ_0 for different approximations to distinguish between strange and nonstrange attractors.

3.3.4 *Different examples*

Let us illustrate the application of this condition using several examples of quasiperiodically forced maps, such as the GOPY-map and the logistic map (cf. Chapter 2). The above mentioned maps are 1-dimensional quasiperiodically forced maps. In this case we check whether the attractor of the quasiperiodically forced system can be represented as smooth curve $x = G(\theta)$, then the attractor is a smooth invariant curve. Otherwise the attractor is strange and consists of a curve which is nowhere differentiable.

We now discuss, how the general approach above can be implemented practically. In fact, we have already demonstrated it for the case of a smooth attractor (system (3.8, 3.9), Fig. 3.1) and for an SNA (system (3.10, 3.11), a series of figures 3.2-3.7). Summarizing the latter case, the k-th approximation of the attractor exhibits bifurcations for any k as the parameter θ_0 changes. For those values of θ_0, for which one of the θ_n is very close to the value 1/4 (where $\cos(2\pi\theta) = 0$) there is only one stable fixed point $x = 0$, while for other values of θ_0 either a pair of stable fixed points or a symmetric period–2 cycle exists. Note, that in the bifurcation points observed here, the derivative of one branch of attracting sets with respect to θ_0 is infinite. All occurring bifurcation points in this example are of such type that the tangent of one branch of attracting sets is orthogonal to the θ_0 axis (turning points, pitchfork bifurcations and period doublings). This picture of bifurcations is qualitatively the same for all k which has been checked for periods up to $q_k = F_k = 987$. Furthermore, the considered interval $0 \leq \theta_0 < 1/F_k$ gets smaller and smaller with increasing k so that the total number of bifurcation points for $\theta_0 \in [0, 1)$ increases. In the limit $k \to \infty$ there are infinitely many points possessing an infinite derivative with respect to θ_0 on smaller and smaller scales. Therefore, we conclude that the limiting attractor cannot be represented by a smooth curve $x = G(\theta)$, *i.e.* it is strange.

Let us now consider the modified GOPY-model (2.15, 2.16), which we rewrite here for convenience

$$x_{n+1} = 2a \tanh x_n \cos 2\pi\theta_n + \alpha \cos 2\pi(\theta_n + \beta) , \tag{3.18}$$

$$\theta_{n+1} = \theta_n + \omega \pmod 1 . \tag{3.19}$$

The parameters $\beta = 0.125$ and $a = 1.5$ are fixed so that the system possesses a strange nonchaotic attractor for $\alpha = 0$. Let us now discuss the case $\alpha \neq 0$. The series of panels in Fig. 3.8 shows different rational approximations for $\alpha = 0.2$. We see that with increasing k the bifurcations disappear. This property is valid for all $\alpha \neq 0$, but for small α the bifurcations are observed for higher-order approximations, see [Pikovsky and Feudel 1995] for details.

While in the example discussed above the possible long-term states appearing in rational approximations are only fixed points and period-2 orbits we would like to demonstrate another example where also chaotic states are possible. Such an example is the forced logistic map (2.27, 2.28), which exhibits also chaotic motion depending on the phase shift θ_0 when an SNA

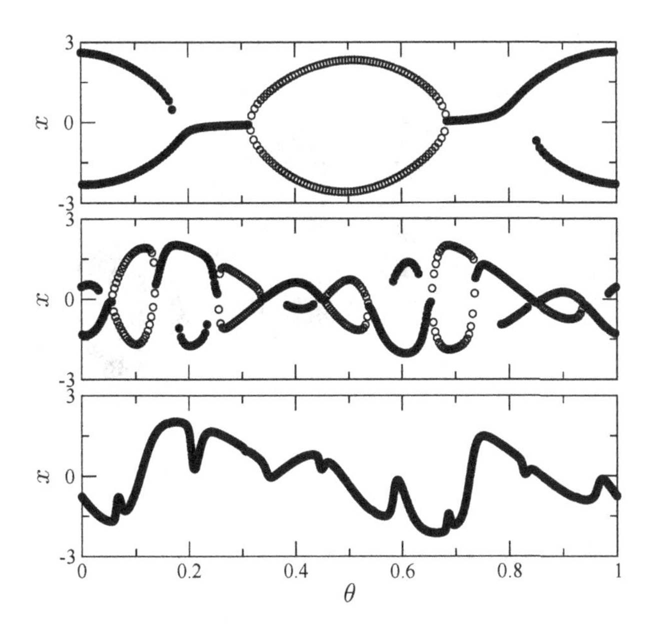

Fig. 3.8 Rational approximations to the attractor in system (3.18, 3.19). Filled circles show stable fixed points, open circles show period-two cycles. Top panel: the first approximation $\omega_1 = 1$. Middle panel: the 4th approximation $\omega_4 = 3/5$. Bottom panel: the 7th approximation $\omega_7 = 13/21$, here only one type of attractor is present and no bifurcations occur.

emerges. Fig. 3.9 shows two rational approximations with $\omega_6 = 8/13$ and $\omega_{12} = 144/233$ for the SNA presented in Fig. 2.7.

Rational approximations are not only a useful tool to analyze the geometrical structure of attractors, but also to study the behavior in parameter space. It is sometimes more convenient to compute bifurcation diagrams using rational approximations instead of the quasiperiodically forced systems directly. Such diagrams give usually already for rather low approximation values k a quite good picture of the dynamics depending on several parameters (see an example Fig. 3.10). Indeed, for every rational approximation it is rather easy to find, whether bifurcations depending on the phase shift occur. If bifurcations are observed, the attractor should be characterized as a strange one. One can also check whether the attractor is chaotic: to accomplish this, one needs to calculate the largest Lyapunov exponent for each value of the phase shift, and then average. If the resulting quantity is positive, the attractor in this approximation should be classified as a chaotic one, otherwise it is nonchaotic. In this way Fig. 3.10 has been obtained.

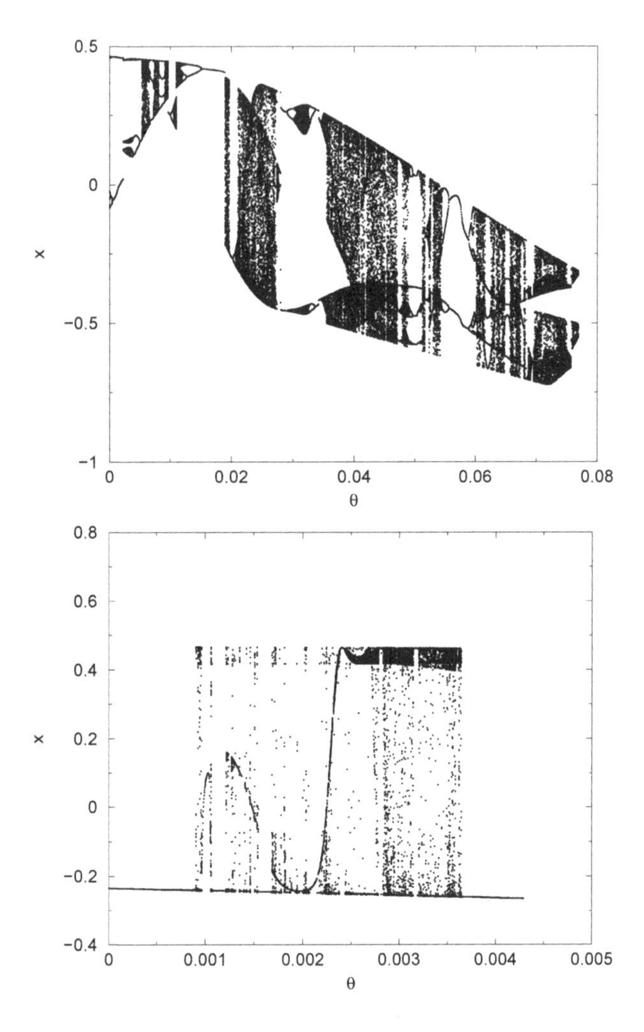

Fig. 3.9 Rational approximations for the SNA in the forced logistic map shown in Fig. 2.7 (represented in the interval $\theta_0 \in [0, 1/q)$). Top panel: $\omega_6 = 8/13$; bottom panel: $\omega_{12} = 144/233$.

3.4 Bibliographic notes

In this chapter we have explained the method of rational approximations following the approach of [Pikovsky and Feudel 1995]; as illustrations mainly 1-dimensional quasiperiodically forced maps have been used. These methods can be extended easily to higher-dimensional maps, as this has been

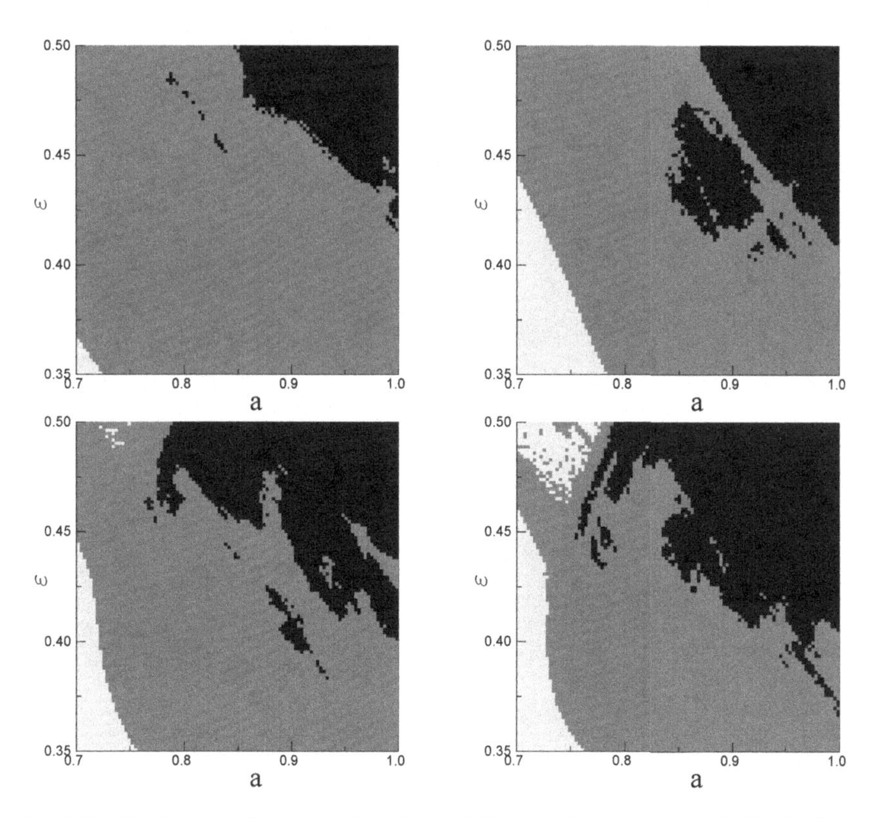

Fig. 3.10 Partition of the parameter plane of the forced logistic map (2.27, 2.28) in different rational approximations. Light gray denotes a smooth torus, gray denotes an SNA and black denotes chaos, like in Fig. 2.3. Top left panel: $\omega_4 = 3/5$, top right panel: $\omega_5 = 5/8$, bottom left panel: $\omega_6 = 8/13$, bottom right panel: $\omega_7 = 13/21$.

done for the quasiperiodically forced Hénon and ring map by Sosnovtseva et al. [1996]. For a treatment of the circle map see [Feudel et al. 1995a]. Kim et al. [2003a] used rational approximations for a skew shift model. Datta et al. [2004] discussed the fractalization transition to strange nonchaotic attractors in terms of rational approximations.

Chapter 4

Stability and Instability

The investigation of stability properties of the motion is the most powerful tool in studies of dynamical systems. Already for distinguishing different types of equilibria and simple periodic motions the notions of Lyapunov stability, asymptotic stability, etc. are extremely important. For complex dynamics the calculation of the Lyapunov exponent is of major significance, as it provides a practical, numerically feasible tool for distinguishing chaotic and regular motions.

In this chapter we describe the characterization of strange nonchaotic attractors via stability properties of the dynamics. We start with the usual Lyapunov exponents. Then the sensitivity to the phase of the forcing is introduced, which is a specific quantity for the characterization of quasiperiodically forced systems. We show how using this quantity one can distinguish SNA from non-strange attractors. Next we demonstrate that the phase sensitivity is closely related to special properties of the finite-time Lyapunov exponents.

4.1 Theoretical consideration

Physically, the stability of the dynamics is characterized by the sensitivity to initial conditions. This sensitivity can be determined for statistically stationary states, e.g. for the motion on an attractor. If this motion demonstrates sensitive dependence on initial conditions, then it is chaotic. In the popular literature this is often called the "Butterfly Effect", after the famous "gedankenexperiment" of Edward Lorenz: if a perturbation of the atmosphere due to a butterfly in Brazil induces a thunderstorm in Texas, then the dynamics of the atmosphere should be considered as an unpredictable and chaotic one. By contrast, stable dependence on initial conditions means that the dynamics is regular. The stability properties of the dynamics are discussed in all standard textbooks on nonlinear dynamics and chaos.

Here we apply these ideas to strange nonchaotic attractors. We start with the basic model of a quasiperiodically forced system

$$x_{n+1} = f(x_n, \theta_n) , \tag{4.1}$$

$$\theta_{n+1} = \theta_n + w . \tag{4.2}$$

Let us look, how a small perturbation of a state x_n, θ_n evolves in time. To this end we need to linearize the system (4.1, 4.2), as a result we get linear equations for the perturbations

$$\delta x_{n+1} = \frac{\partial f(x_n, \theta_n)}{\partial x_n} \delta x_n + \frac{\partial f(x_n, \theta_n)}{\partial \theta_n} \delta \theta_n , \tag{4.3}$$

$$\delta \theta_{n+1} = \delta \theta_n . \tag{4.4}$$

The system (4.3, 4.4) has one peculiarity which is characteristic for all situations where one subsystem is driving another one: the perturbations of θ are not influenced by the perturbations of x. As a result, by iterating (4.3, 4.4) we obtain for the evolution of the perturbations after n time steps

$$\begin{pmatrix} \delta x_n \\ \delta \theta_n \end{pmatrix} = \begin{pmatrix} \frac{\partial x_n}{\partial x_0} & \frac{\partial x_n}{\partial \theta_0} \\ 0 & 1 \end{pmatrix} \begin{pmatrix} \delta x_0 \\ \delta \theta_0 \end{pmatrix} . \tag{4.5}$$

The components of the Jacobian matrix in (4.5) satisfy the following recursions

$$\frac{\partial x_n}{\partial x_0} = \frac{\partial f(x_{n-1}, \theta_{n-1})}{\partial x_{n-1}} \frac{\partial x_{n-1}}{\partial x_0} = \prod_{k=0}^{n-1} \frac{\partial f(x_k, \theta_k)}{\partial x_k} , \tag{4.6}$$

$$\frac{\partial x_n}{\partial \theta_0} = \frac{\partial f(x_{n-1}, \theta_{n-1})}{\partial x_{n-1}} \frac{\partial x_{n-1}}{\partial \theta_0} + \frac{\partial f(x_{n-1}, \theta_{n-1})}{\partial \theta_{n-1}} . \tag{4.7}$$

The quantity $\frac{\partial x_n}{\partial x_0}$ determines the sensitivity of the variable x to initial conditions, its asymptotic growth/decay rate is nothing else than the Lyapunov exponent:

$$\lambda = \lim_{n \to \infty} \frac{\ln \left| \frac{\partial x_n}{\partial x_0} \right|}{n} = \lim_{n \to \infty} \frac{1}{n} \sum_{k=0}^{n-1} \ln \left| \frac{\partial f(x_k, \theta_k)}{\partial x_k} \right| . \tag{4.8}$$

For SNA the Lyapunov exponent is negative, this means the absence of sensitivity to initial conditions. Thus the derivative

$$\left| \frac{\partial x_n}{\partial x_0} \right| \sim e^{\lambda n} \tag{4.9}$$

is exponentially small for large n, and the corresponding component of the Jacobian matrix tends to zero.

The only nontrivial component in the Jacobian matrix in (4.5) is the derivative $\frac{\partial x_n}{\partial \theta_0}$. This quantity describes the dependence of the driven variable x on the driving phase θ. Because $\frac{\partial \theta_n}{\partial \theta_0} = 1$, we can drop the index and will call $\frac{\partial x_n}{\partial \theta}$ the *phase sensitivity*. This quantity obeys the recursion relation (4.7), which, contrary to (4.6), is nonhomogeneous. Thus, the general formula for the phase sensitivity is more complex than (4.6).

Iterations of (4.7) yield

$$\begin{aligned} \frac{\partial x_n}{\partial \theta_0} = & \frac{\partial f(x_{n-1}, \theta_{n-1})}{\partial \theta_{n-1}} \\ & + \frac{\partial f(x_{n-2}, \theta_{n-2})}{\partial \theta_{n-2}} \frac{\partial f(x_{n-1}, \theta_{n-1})}{\partial x_{n-1}} \\ & + \frac{\partial f(x_{n-3}, \theta_{n-3})}{\partial \theta_{n-3}} \frac{\partial f(x_{n-1}, \theta_{n-1})}{\partial x_{n-1}} \frac{\partial f(x_{n-2}, \theta_{n-2})}{\partial x_{n-2}} \\ & + \frac{\partial f(x_{n-4}, \theta_{n-4})}{\partial \theta_{n-4}} \frac{\partial f(x_{n-1}, \theta_{n-1})}{\partial x_{n-1}} \frac{\partial f(x_{n-2}, \theta_{n-2})}{\partial x_{n-2}} \frac{\partial f(x_{n-3}, \theta_{n-3})}{\partial x_{n-3}} \\ & + \dots \end{aligned} \tag{4.10}$$

We can write this series as

$$\frac{\partial x_n}{\partial \theta} = \sum_{k=0}^{n-1} \frac{\partial f(x_k, \theta_k)}{\partial \theta_k} X_{k+1,n} \,, \tag{4.11}$$

where

$$X_{k+1,n} = \prod_{j=k+1}^{n-1} \frac{\partial f(x_j, \theta_j)}{\partial x_j} = \frac{\partial x_n}{\partial x_{k+1}} \,. \tag{4.12}$$

Note that the factors $X_{k,n} = \frac{\partial x_n}{\partial x_k}$ in (4.11) are the same quantities as defined in (4.6), i.e. they measure the sensitivity to the initial condition. However, contrary to the discussion of the Lyapunov exponent above, we cannot conclude that the phase sensitivity is small: due to summation in (4.11) it is at least finite, and can be even diverging. Indeed, for a smooth dynamical system we can assume that the derivatives $\frac{\partial f(x_k,\theta_k)}{\partial \theta_k}$ are bounded.

The convergence of the sum in (4.11) then depends on the behavior of the quantities $\frac{\partial x_n}{\partial x_k}$. For large $(n-k)$ the asymptotics (4.9) is valid and these quantities are small, however we need to estimate $\frac{\partial x_n}{\partial x_k}$ for finite $(n-k)$ as well. This can be accomplished by introducing so-called finite-time Lyapunov exponents (also called local Lyapunov exponents) according to

$$\Lambda_n(x_0) = \frac{\ln|\frac{\partial x_n}{\partial x_0}|}{n} = \frac{1}{n}\sum_{k=0}^{n-1} \ln\left|\frac{\partial f(x_k, \theta_k)}{\partial x_k}\right| . \tag{4.13}$$

Note that the difference to (4.8) is only in taking the limit, thus the usual Lyapunov exponent is the limit of the finite-time ones

$$\lambda = \lim_{n\to\infty} \Lambda_n . \tag{4.14}$$

Of main interest for us are the factors in (4.11) which can be represented as

$$\frac{\partial x_n}{\partial x_k} = \prod_{j=k}^{n-1} \frac{\partial f(x_j, \theta_j)}{\partial x_j} = \pm\exp[(n-k)\Lambda_{n-k}(x_k)] , \tag{4.15}$$

to estimate them we need to know the distribution of the finite-time Lyapunov exponents. The form of this distribution can be guessed from the theory of large deviations. If one assumes that (4.13) is a sum of weakly correlated quantities, then one can asymptotically write

$$Prob(\Lambda_n) \sim \exp[n\phi(\Lambda)] , \tag{4.16}$$

where $\phi(\Lambda)$ is a scaling function giving the asymptotic form of the distribution. One can see that for large n the distribution of Λ_n is concentrated near the maximum of $\phi(\Lambda)$, thus this maximum is exactly the Lyapunov exponent λ.

We now discuss possible types of the function $\phi(\Lambda)$. In general, we expect that it is non-zero in some interval $\Lambda_{min} < \Lambda < \Lambda_{max}$. Suppose first that $\Lambda_{max} < 0$ (see Fig. 4.1a). Then we can estimate

$$\left|\frac{\partial x_n}{\partial x_k}\right| \sim \exp[(n-k)\Lambda_{n-k}(x_k)] < \exp[(n-k)\Lambda_{max}] . \tag{4.17}$$

Therefore, one can impose an exponential estimate on the terms in the sum (4.11), these terms are bounded from above by $const \cdot e^{(n-k)\Lambda_{max}}$. Since $\Lambda_{max} < 0$, this sum converges, what means that the phase sensitivity is finite. The bound of the derivative $\frac{\partial x}{\partial \theta}$ yields that the dependence of x on

θ is a relatively smooth curve (or a set of curves), i.e. we have a smooth torus as an attractor in the system.

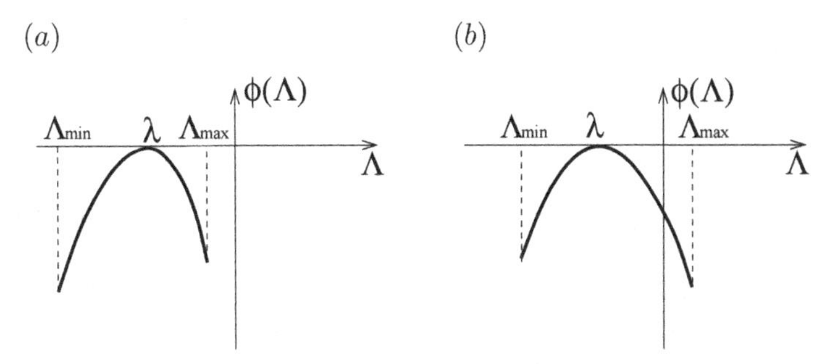

Fig. 4.1 A sketch of the scaling function for the distribution of finite-time Lyapunov exponents. In both cases the average exponent λ is negative; in case (a) positive finite-time exponents are unprobable, while in case (b) they can occur.

Another type of behavior can be expected if $\Lambda_{max} > 0$ (see Fig. 4.1b). In this case one cannot bound the derivative $\frac{\partial x_n}{\partial x_k}$ from above: large values of this quantity are possible, although rare. Thus, in the sum (4.11) these large values dominate, and the sum may diverge. Let us estimate the largest term in the sum (4.11). A piece of trajectory of length m with a finite-time exponent Λ can be observed within a long observation time interval n if its probability to occur is of order $1/n$, i.e. if $\exp[m\phi(\Lambda)] \sim 1/n$. From this we get $m \sim -\frac{\ln n}{\phi(\Lambda)}$. The corresponding value of the derivative $\frac{\partial x_n}{\partial x_k}$ is

$$\exp[m\Lambda] = \exp\left[-\ln n \frac{\Lambda}{\phi(\Lambda)}\right] = n^{-\frac{\Lambda}{\phi(\Lambda)}} . \tag{4.18}$$

Thus, the maximal value of the derivative that can be achieved over the time interval n is

$$\max_{0<k<n} \left| \frac{\partial x_k}{\partial \theta} \right| \sim n^{\mu} , \qquad \text{where} \qquad \mu = \max_{\Lambda>0} \left(-\frac{\Lambda}{\phi(\Lambda)} \right) . \tag{4.19}$$

The quantity μ is positive, thus for large n very large contributions to the sum (4.11) can occur. As a result, we come to the conclusion that the dependence of x on θ cannot be a smooth curve or a set of curves, and the attractor is strange.

Summarizing the discussion above, we can conclude that the existence of SNA can be derived from the stability properties of the dynamics. An SNA exists if the asymptotic distribution of finite-time Lyapunov exponents is like in Fig. 4.1b, i.e. there is a non-zero probability for a positive finite-time exponent to occur. Correspondingly, the phase sensitivity for the SNA is unbounded. In the case of a smooth attractor, the distribution of finite-time Lyapunov exponents is restricted (for large times) to the negative domain of Λ, and the phase sensitivity is bounded.

The consideration above is heavily based on the assumption that one can describe the distribution of finite-time Lyapunov exponents statistically and write (4.16). This already presumes some degree of irregularity in the dynamics. For regular dynamics one in fact expects that the distribution is asymptotically a delta-function, i.e. in Fig. 4.1 $\Lambda_{min} = \lambda = \Lambda_{max}$. But if, e.g., one considers a stable periodic motion, then there can be pieces of the limit cycle with different degree of stability, or even unstable pieces. However, when considering large time intervals (larger than the cycle period), all unstable parts are compensated by stable ones and one cannot speak of positive finite-time exponents.

The situation changes in quasiperiodically driven systems. Here the attractor is not represented by just one trajectory, like in the case of a limit cycle, but by a set of trajectories. The condition that there can be positive finite-time Lyapunov exponents means that there are some, rather unprobable, trajectories that are unstable. For several SNAs these unstable orbits can be identified, since they become embedded in the attractor during the emergence of the SNA (cf. Chapter 6), but in many cases only a statistical numerical treatment is possible.

4.2 Numerical examples

4.2.1 *Discrete time mappings*

For discrete-time one-dimensional mappings we have just to follow the theoretical framework outlined above. As an example we consider the generalized GOPY-model (2.15, 2.16)

$$x_{n+1} = 2a \tanh(x_n) \cos(2\pi\theta_n) + \alpha \cos(2\pi(\theta_n + \beta)) , \tag{4.20}$$

$$\theta_{n+1} = \theta_n + \omega . \tag{4.21}$$

For $\alpha = 0$ it exhibits an SNA, and for $\alpha \neq 0$ a smooth torus, see Sections 2.2, 3.3.4.

We first demonstrate that in the case of a non-fractal attractor the derivative can be calculated with formula (4.7). The results are shown in Fig. 4.2. One can see that the iterations of (4.7) indeed give the correct

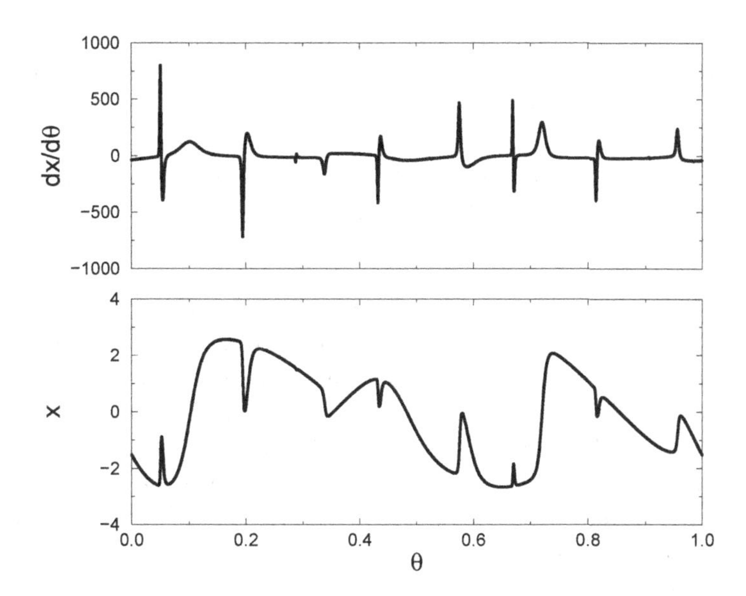

Fig. 4.2 The attractor in system (4.20) for $a = 1.5$, $\beta = 1/8$, $\alpha = 0.4$ (bottom panel) and the derivative $\frac{\partial x}{\partial \theta}$ (upper panel) The latter has been obtained by iterations of Eq. (4.7). Although the derivative attains large values, it remains bounded.

values of the phase derivative on the attractor, at some points this derivative is quite large but finite.

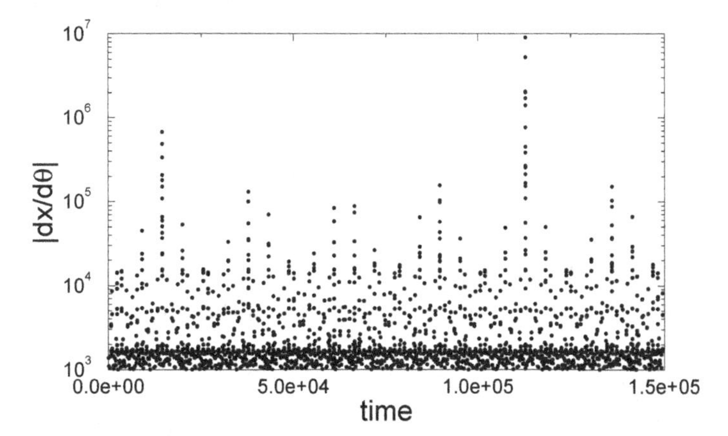

Fig. 4.3 The time series of $\left|\frac{\partial x}{\partial \theta}\right|$ obtained by iteration of (4.7) for system (4.20,4.21) $a = 1.5$ and $\alpha = 0$. Only values exceeding 10^3 are depicted; one can see that sometimes very large values of the derivative are observed.

Another picture is observed when the relation (4.7) is used for calculating the phase derivative for an SNA. Here the calculation gives the results presented in Fig. 4.3. The time dependence of the calculated derivative $|\frac{\partial x_n}{\partial \theta_0}|$ is highly intermittent, and at some instants of time its values are extremely large. Because the large values of the derivative occur relatively rare, it is convenient to introduce the quantity

$$\gamma(n, x, \theta) = \max_{0 \le k \le n} \left| \frac{\partial x_k}{\partial \theta} \right| , \tag{4.22}$$

where we write explicitly its dependence on the initial point x, θ. This quantity is nothing else but the "envelope" of the graph in Fig. 4.3. It is convenient to draw γ as a function of n in logarithmic scale, this dependence is shown in Fig. 4.4. In fact, we show here this quantity for different initial conditions. Although for different initial points the behavior of $\gamma(n)$ is similar, one can hardly simply average over initial conditions (like in the case of the Lyapunov exponent) because due to intermittency a very large value of γ can occur already for small n. A more suitable characteristics appears to be the minimum over initial points

$$\Gamma(n) = \min_{x,\theta} \gamma(n, x, \theta) . \tag{4.23}$$

This quantity (see Fig. 4.5) can be used for characterizing the stability of the dynamics with respect to the phase of the external force. It can be roughly compared with the theoretical estimate Eq. (4.19). For an SNA typically $\Gamma(n) \sim n^{\mu}$, and the quantity μ is called *phase sensitivity exponent.* For a smooth torus $\Gamma(n)$ does not grow and $\mu = 0$.

The next numerical results illustrate the validity of the assumption (4.16) and Fig. 4.1. In Fig. 4.6 we show the distribution of the finite-time Lyapunov exponents for the system under consideration. One can see a qualitative agreement with Fig. 4.1(b), although the n-dependence is still present in Fig. 4.6 due to slow convergence to the asymptotics assumed in Eq. (4.16). The value of Λ_{max} can be easily estimated from Eq. (4.20). When $\alpha = 0$, the curve $x = 0$ is invariant and on this curve

$$\left\langle \ln \left| \frac{dx_{n+1}}{dx_n} \right| \right\rangle = \ln(2|a|) + \ln\langle |\cos 2\pi\theta| \rangle = \ln |a| . \tag{4.24}$$

One can see that this invariant curve belongs to the SNA, thus a trajectory comes arbitraryly close to it. During these visits the maximal value of possible finite-time Lyapunov exponents is achieved, for the parameters chosen $\Lambda_{max} = \ln(1.5) \approx 0.405$ in agreement with Fig. 4.6.

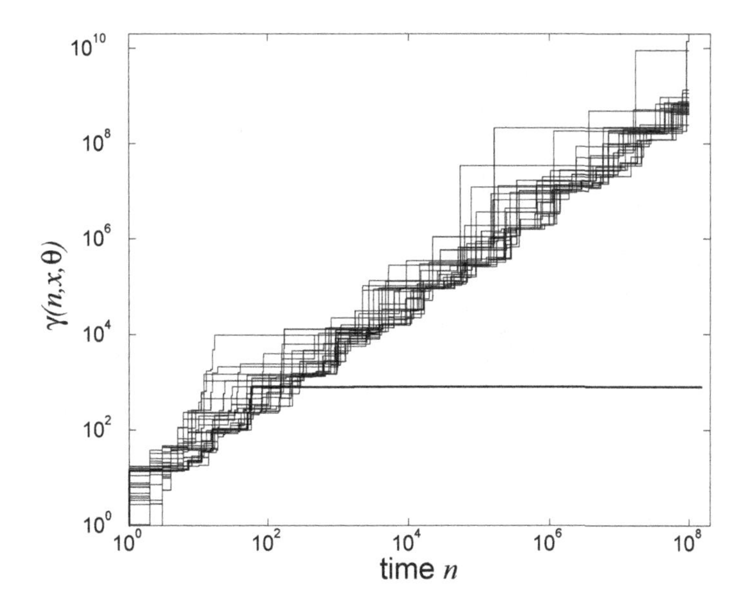

Fig. 4.4 The envelope of the derivative $|\frac{\partial x_n}{\partial \theta_0}|$, calculated for the same system as in Fig. 4.3 for different initial conditions. For comparison we show also the envelope for the non-strange attractor shown in Fig. 4.2 with the bold line.

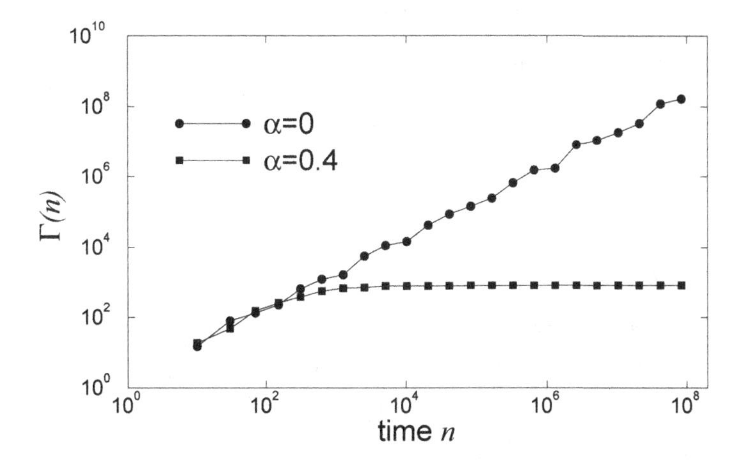

Fig. 4.5 The characterization of the phase sensitivity according to (4.23) for system (4.20,4.21) for two values of parameter α, corresponding to the cases of SNA ($\alpha = 0$) and smooth torus ($\alpha = 0.4$). Γ grows nearly linearly with n in the case of SNA.

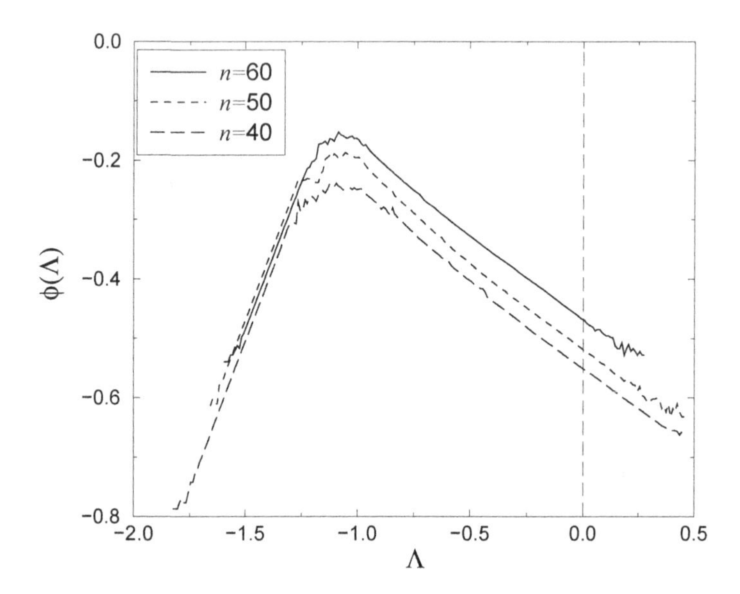

Fig. 4.6 The scaling function for the distribution of finite-time Lyapunov exponents in system (4.20,4.21) for $\alpha = 0$, $a = 1.5$. Three estimations of the scaling function according to (4.16) for different values of n are shown.

4.2.2 *Continuous time systems*

A generalization of the theory above to continuous-time systems is straightforward. Consider a system of N differential equations driven by M incommensurate frequencies

$$\frac{dx_l}{dt} = F_l(x_1, \ldots, x_N, \phi_1, \ldots, \phi_M) , \qquad l = 1, \ldots, N , \tag{4.25}$$

$$\frac{d\phi_k}{dt} = \omega_k , \qquad k = 1, \ldots, M . \tag{4.26}$$

Then the derivatives of the variables x_l with respect to the phases $y_{lk} = \frac{\partial x_l}{\partial \phi_k}$ are governed by the inhomogeneous system

$$\frac{dy_{lk}}{dt} = \sum_j \frac{\partial F_l}{\partial x_j} y_{jk} + \frac{\partial F_l}{\partial \phi_k} . \tag{4.27}$$

Solving (4.27) one gets the same quantity as iterating Eq. (4.11). Accordingly, the quantities defined by Eqs. (4.22),(4.23) can be straightforwardly calculated for a continuous-time system as well. We illustrate in Fig. 4.7 the approach described with the calculation of the quantity $\Gamma(t)$ for the simplest one-dimensional system demonstrating SNA (cf. (2.33)):

$$\begin{aligned}\frac{dx}{dt} &= -\sin(2\pi x) + b_1 \sin(2\pi\phi_1) + b_2 \sin(2\pi\phi_2) \;, \\ \frac{d\phi_1}{dt} &= \omega_1 \;, \\ \frac{d\phi_2}{dt} &= \omega_2 \;.\end{aligned} \tag{4.28}$$

Both derivatives $|\frac{\partial x}{\partial \phi_1}|$ and $|\frac{\partial x}{\partial \phi_2}|$ grow unboundedly in time for the case of SNA and saturate for the smooth torus.

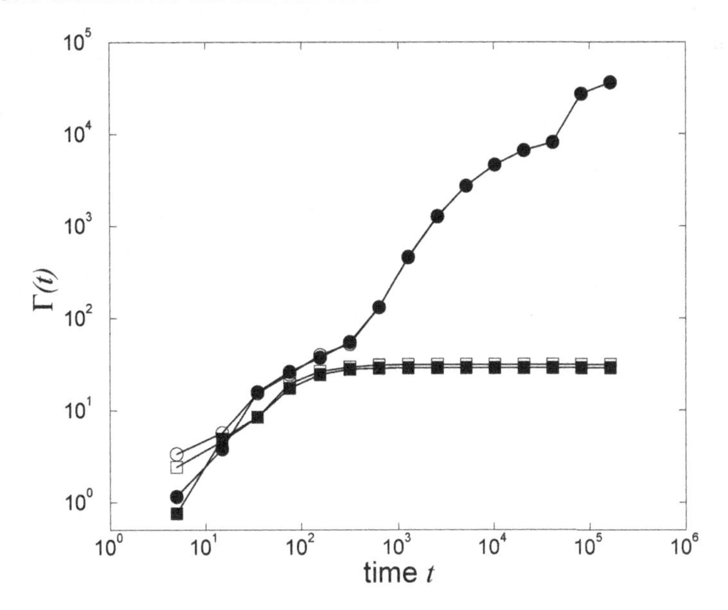

Fig. 4.7 Time dependence of the phase sensitivity in model (4.28) for $\omega_1 = 2\pi$, $\omega_2 = (\sqrt{5}-1)\pi$, and two amplitudes of forcing: circles correspond to the case of SNA $b_1 = b_2 = 4\pi$, while boxes correspond to a smooth torus at $b_1 = b_2 = 0.7\pi$. Open and filled symbols mark derivatives with respect to ϕ_1 and ϕ_2, respectively; they are practically undistiguishable.

4.3 Bibliographic notes

In this chapter we follow mainly the paper [Pikovsky and Feudel 1995] where the characterization of the phase sensitivity of SNAs has been introduced. For a treatment of the circle map see [Feudel et al. 1995a]. In [Stark

1997] it has been proven that if the stability of a quasiperiodically driven system is uniform, i.e. there are no unstable pieces of trajectories, then the attractor is a smooth graph; this means that some degree of instability must be present for an SNA to be observed albeit the largest Lyapunov exponent is negative. The concept of the phase sensitivity exponent can be generalized to higher-dimensional maps. This has been illustrated by applications to the quasiperiodically forced Hénon and ring map in [Sosnovtseva et al. 1996]. Kapitaniak [1995] calculated the distribution of finite-time Lyapunov exponents for a quasiperiodically forced pendulum. Prasad and Ramaswamy [1999] studied the form of the distribution of finite-time Lyapunov exponents for different types of chaos and nonchaotic behavior. Shuai et al. [2001] have demonstrated that attempts to reconstruct the Lyapunov exponent from a time series lead to positive values of the exponent. Fu et al. [2002] used the concept of phase sensitivity to characterize the dynamics near the TDT point (cf. Section 7.6). Shuai and Wong [1998]; Shuai and Durand [2000] studied variations of finite-time Lyapunov exponents under low-frequency quasiperiodic forcing. For a similar low-frequency forcing Shi et al. [2001] have demonstrated that the finite computer precision may lead to spurious chaos in such a situation. Yu et al. [1999a] considered an optical system driven with a low-frequency force and focused on finite-time Lyapunov exponents.

Chapter 5

Fractal and statistical properties

For chaotic dynamical systems it is quite natural to treat the deterministic processes as random ones, and to describe them statistically. This is the subject of ergodic theory. It happens that many tools developed for random processes can be applied to deterministic time series as well, e.g. the spectral analysis, or calculation of the diffusion constant. One can apply statistical tools also to regular dynamical systems, here many statistical characteristics become degenerate or trivial. For a periodic process, e.g., the correlation function is purely periodic and the power spectrum is discrete. Strange nonchaotic attractors lie in between, and their statistical properties are rather nontrivial. In this chapter we first describe the properties of the invariant measure of SNAs. Here the main question is the fractality, we argue that at least in some cases an SNA is a multifractal object. Next, we discuss properties of correlations and spectra. We demonstrate, that there is an example of SNA with a singular continuous (fractal) spectrum.

5.1 Fractal properties of SNA

Already the name of strange nonchaotic attractors suggests that these objects are fractals in phase space. However, a direct numerical analysis of the corresponding fractal properties is extremely difficult, and only a few examples exist where the fractality can be shown analytically. As the simplest attractor in a quasiperiodically forced system is a torus whose dimension is one in the case of a map, the dimension of an SNA cannot be less than one. One can easily estimate the Lyapunov dimension of an SNA: because the Lyapunov exponents on it are negative, the Lyapunov dimension calculated according to the Kaplan-Yorke formula (see, e.g., [Ott 1992]) is one.

In some cases it is possible to give a precise estimate for other generalized dimensions.

We first remind how generalized dimensions are defined. Suppose an attractor is covered with boxes B of size ϵ. If the invariant measure on the attractor is μ, each box has measure $\mu(B)$. Then the so-called partition sum $Z_\epsilon(q)$ is calculated

$$Z_\epsilon(q) = \sum_{\mu(B)\neq 0} [\mu(B)]^q \tag{5.1}$$

and the generalized dimensions are defined by the limit

$$D_q = \frac{1}{q-1} \lim_{\epsilon\to 0} \frac{\log Z_\epsilon(q)}{\log \epsilon} . \tag{5.2}$$

Practically, if the distribution μ is approximated by a sample of N points, then $\mu(B) \approx \frac{n_i}{N}$ where n_i is the occupation of the i-th box. Of primary interest are dimensions with $q = 0, 1, 2$. The box-counting or capacity dimension D_0 is calculated according to

$$D_0 = -\lim_{\epsilon\to 0} \frac{\log M(\epsilon)}{\log \epsilon} , \tag{5.3}$$

where $M(\epsilon)$ is the number of non-empty boxes. For the information dimension D_1 we get

$$D_1 = \lim_{\epsilon\to 0} \frac{\sum_i \frac{n_i}{N} \log \frac{n_i}{N}}{\log \epsilon} . \tag{5.4}$$

Finally, the correlation dimension D_2 reads

$$D_2 = \lim_{\epsilon\to 0} \frac{\log \sum_i \left(\frac{n_i}{N}\right)^2}{\log \epsilon} . \tag{5.5}$$

¿From the definitions (5.3)-(5.5) one can see that the box-counting dimension depends only on the topological properties of an attractor, while D_1 and D_2 depend also on the probability distribution on it. They obey an inequality $D_2 \leq D_1 \leq D_0$. It is possible that all three dimensions are different, this case is called multifractal. Notice that the Lyapunov dimension for a large class of dynamical systems coincides with the information dimension. Although this result may not hold for SNA, it indicates that $D_1 = 1$ might be a good guess.

The box-counting dimension can be estimated in some particular cases. Following Hunt and Ott [2001], we consider the following solvable example:

$$x_{n+1} = x_n + \theta_n + \varepsilon\psi(2\pi x_n) \pmod 1 , \tag{5.6}$$

$$\theta_{n+1} = \theta_n + \omega \pmod 1 . \tag{5.7}$$

Here $\psi(2\pi x)$ is a 2π-periodic function with $|\varepsilon\psi'| < (2\pi)^{-1}$, so that the mapping (5.6) is one-to-one. In the case $\varepsilon = 0$ system (5.6, 5.7) is the skew shift, described in detail in [Cornfeld et al. 1982]. The skew shift is an area-preserving map and covers the torus $0 \leq \theta, x < 2\pi$ uniformly. This means that all dimensions of the skew shift are equal to two. For $\varepsilon \neq 0$ the Lyapunov exponent in (5.6) is generally nonzero. Clearly, it cannot be positive as the mapping (5.6) is non-expanding. Numerics shows that the Lyapunov exponent is negative. If we use the uniform distribution of x for the calculation of λ for small ε, then it follows that the Lyapunov exponent is negative:

$$\lambda = \left\langle \ln\left(1 + \varepsilon\frac{d\psi}{dx}\right)\right\rangle \approx -\varepsilon^2 \left\langle \left(\frac{d\psi}{dx}\right)^2 \right\rangle < 0 . \tag{5.8}$$

It can be argued that the attractor for $\varepsilon \neq 0$ is everywhere dense on the torus, this follows from the fact that any line initially wrapping the torus only in θ-direction, under iterations of (5.6, 5.7) wraps the torus more and more times in x-direction (see the detailed argument by Hunt and Ott [2001]). Therefore, the box-counting dimension of the SNA in system (5.6, 5.7) is two. This proves the multifractal nature of SNA in this particular example.

A convincing, although not exact, argument can be used for the GOPY-model. It is based on rational approximations (see Chapter 3). Some approximations are depicted in Figs. 3.2-3.7. To use these approximations for calculating the quantity $M(\epsilon)$ one associates the scale $1/F_k$ with a particular rational approximation $\omega_k = F_{k-1}/F_k$ and takes this scale as the box size ϵ_k. Then one increases k and in this way the limit $\epsilon \to 0$ in (5.3) is taken. Now if we turn to Fig. 3.7, which shows an approximation with ω_7, one can see that the vertical scale in this Figure is of order 1 while the scale of the θ_0-axis is of order $1/F_k = \epsilon_k$. Thus one has to cover the graph in Fig. 3.7 with boxes of size $\epsilon_k = 1/21$ which is the total width of this graph. One can see that the graph in Fig. 3.7 contains vertical lines, therefore the number of boxes is $\approx (x_{max} - x_{min})/\epsilon_k$. Noting that Fig. 3.7 represents only a part of the full graph Fig. 3.6, and the number of such parts is $F_k = \epsilon_k^{-1}$, we conclude that the number of boxes is $M(\epsilon) \approx (x_{max} - x_{min})\epsilon_k^{-2}$. Substituting this in (5.3) yields $D_0 = 2$.

The consideration above is based on the fact that in any rational approximation the attractor looks roughly as in Fig. 3.7: it consists of two horizontal lines (at $x \approx \pm 1$) coupled with vertical crosspieces. This indeed is the case for any k, as there always exist such phase shifts θ_0 that for some m $\theta_0 + m\omega_k = \pi/2$ so that $\cos(\theta_0 + m\omega_k)$ and variable x for this θ_0, calculated according to (3.10, 3.11) vanishes. These are the points of crosspieces.

Quantitatively, the widths of crosspieces decrease rapidly for large k, so that it is rather difficult to obtain the picture like Fig. 3.7 numerically for $k > 10$. This means that if we take a finite set of points θ_0 and choose them randomly (or, say, equidistant), we may miss the vertical crosspieces in a high-order rational approximation of the attractor, so that only two horizontal lines at $x \approx \pm 1$ will be seen. Then the number of boxes to cover these horizontal lines is of order 1, the total number of boxes is $M \sim \epsilon^{-1}$ and one obtains the box counting dimension close to one. This is the reason why it is extremely difficult, if possible at all, to study dimensions of SNAs with direct numerical algorithms.

The consideration above can be reformulated in terms of the invariant measure. Although the k-th approximation to the attractor consists of horizontal lines and vertical crosspieces, the measure is concentrated on horizontal lines and the crosspieces have a measure that vanishes in the limit $k \to \infty$. This means that for calculations of dimensions D_1 and D_2, where the boxes are weighted with their measure (see (5.4),(5.5)), only the horizontal lines contribute. Therefore, one expects that $D_1 \approx D_2 \approx 1$, although this conclusion is very hard to verify numerically, see [Ding et al. 1989a].

5.2 Correlations and spectra of SNA

5.2.1 *Power spectra of regular and irregular motions*

We start with recalling the usual definitions of the correlations and spectra. Thereby we concentrate on the case of discrete time. Given a stationary time series x_n (hereafter we assume that $\langle x \rangle = 0$) the non-normalized autocorrelation function is defined as

$$C(\tau) = \langle x_n x_{n+\tau} \rangle \ . \tag{5.9}$$

For a periodic process $C(\tau)$ is periodic with the period T: $C(0) = C(T) = \langle x^2 \rangle$.

For a quasiperiodic process the autocorrelation function is also quasiperiodic. Indeed, suppose that the process x_n can be represented as

$$x_n = G(\theta_n) \,, \tag{5.10}$$

where $G(\theta) = G(\theta + 1)$ is a periodic function with period one and $\theta_n = \omega n + \theta_0$. This representation appears naturally if x_n is a smooth torus in a mapping of type (4.1, 4.2). In the calculation of the correlation function one has to average over the initial phase θ_0. Writing G as a Fourier series

$$G(\theta) = \sum_{k=-\infty}^{\infty} g_k e^{i2\pi k\theta} \,, \qquad g_{-k} = g_k^* \,, \tag{5.11}$$

we obtain after a simple calculation

$$C(\tau) = \int_0^1 d\theta_0 G(\theta_0) G(\theta_0 + \omega\tau) = \sum_{k=-\infty}^{\infty} |g_k|^2 e^{-i2\pi k\tau\omega} \,. \tag{5.12}$$

¿From this formula it follows that the correlation function is a quasiperiodic function of time. In particular, if an irrational ω can be well approximated by a rational number $\frac{p}{q}$, then $C(q) \approx C(0)$, i.e. the correlation function, although being not periodic, attains regularly values close to $C(0)$.

For a chaotic process x_n one expects a decaying correlation function, typically such a decay is exponential in time $C(\tau) \sim \exp[-\kappa\tau]$. We will see in Section 5.2.3 below, that for SNA it is possible to observe a correlation function that neither decays to zero, nor is periodic or quasiperiodic.

Spectral characteristics of the process can be defined in two ways. On one hand one can define the Fourier transform of the time series $\{x_n\}$ at frequency Ω as

$$s(\Omega, T) = \sum_{n=0}^{T-1} x_n e^{i2\pi\Omega n} \,. \tag{5.13}$$

Then the power spectrum component at this frequency is an ensemble average of the properly normalized Fourier transform:

$$S(\Omega) = \lim_{T\to\infty} \frac{\langle |s(\Omega, T)|^2 \rangle}{T} \,. \tag{5.14}$$

(This normalization follows because for a random process we expect the sum in (5.13) to grow diffusively, i.e. $\sim \sqrt{T}$.) Alternatively, one can use the Wiener-Khinchin relation to define the power spectrum as a Fourier transform of the autocorrelation function

$$S(\Omega) = \sum_{\tau=-\infty}^{\infty} C(\tau) \cos 2\pi\Omega\tau \,. \tag{5.15}$$

Clearly, the spectrum of a periodic process contains a finite number of delta-peaks at $\Omega = m/T$. Substituting (5.12) in (5.15) one can easily see that the spectrum contains generally an infinite number of delta-peaks:

$$S(\Omega) = \sum_{k=-\infty}^{\infty} \frac{|g_k|^2}{2} \delta(\Omega - [k\omega]]) \,, \tag{5.16}$$

where $[k\omega]$ means reduction to the interval $-0.5 \leq \Omega \leq 0.5$ usual for the spectrum of discrete processes. For a chaotic process with a decaying correlation the sum (5.15) converges and one has a continuous spectrum with a finite spectral density $S(\Omega)$.

The classification above is widely used to characterize processes and to determine whether they are regular or chaotic. This approach is especially convenient in experiments, because there, contrary to numerical simulations, one can hardly calculate such characteristics of the dynamics as Lyapunov exponents. However, the characterization of the processes according to spectra and correlations has certain drawbacks. First, it is important to notice that there is a power spectrum of a process x_n, but there is no power spectrum of a dynamical system: the observed power spectrum depends on the observable (in particular, in the case of multidimensional dynamics any variable can be used as an observable, or even any scalar function of the variables). Usually this nonuniversality of correlations and spectra is not so important, because one looks on qualitative properties, like the continuity of the spectrum and the periodicity of correlations.

The second important feature is that the spectrum and the correlation function can be a mixture of different types, e. g., the spectrum can be a mixture of a discrete and a continuous one. This happens typically when one observes chaos in a periodically or quasiperiodically driven system, in this case the spectrum can be presented as a superposition of a discrete periodic or quasiperiodic component that follows the driving and a continuous component corresponding to chaos.

5.2.2 *Spectral properties of fractal tori*

The simplest nontrivial situation relevant for SNAs appears when an SNA can be represented by a function (5.10), but this function is non-smooth (fractal). For a smooth function $G(\theta)$ we expect that the Fourier coefficients g_k decay exponentially: $|g_k| \sim \exp[-\text{const} \cdot k]$. For a non-smooth curve the coefficients g_k are expected to decay as a power law.[1] To characterize these two cases quantitatively, one can introduce the quantity

$$N(\sigma) = \sum_{k:\ |g_k|^2 > \sigma} 1 \tag{5.17}$$

which is simply the number of Fourier harmonics with intensities exceeding σ. For a smooth torus

$$N(\sigma) \sim -\ln\sigma\ , \tag{5.18}$$

while for a fractal one

$$N(\sigma) \sim \sigma^{-\alpha} \tag{5.19}$$

with some nontrivial α.

We illustrate this characterization by examining the correlation properties of SNA in Figs. 5.1-5.3. In Fig. 5.1 we depict the spectral characteristics of the smooth torus presented in Fig. 4.2. One can see that the Fourier coefficients decay exponentially and the law (5.18) is valid.

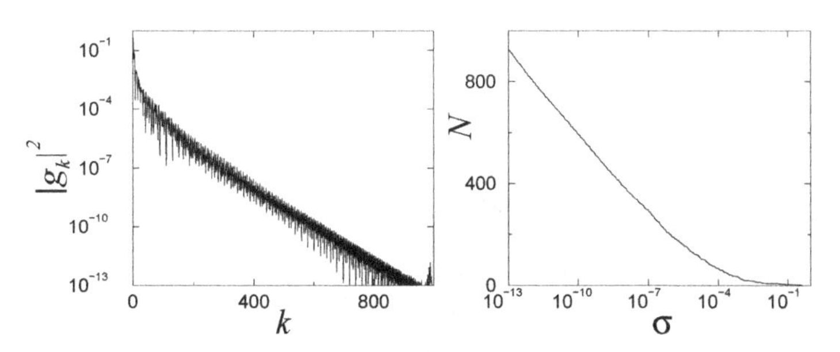

Fig. 5.1 Spectral characteristics of the smooth torus shown in Fig. 4.2 for the observable x_n. Left panel: Fourier coefficients; right panel: the number of Fourier harmonics $N(\sigma)$ that exceed the level σ.

[1] To see this one has to differentiate (5.11) to get $G^{(n)}(\theta) = \sum g_k (i2\pi k)^n e^{i2\pi k\theta}$. For exponentially decaying Fourier coefficients this series converges for any n, while if g_k decays as a power law, the series diverges for some n which means non-smoothness.

In Fig. 5.2 the critical torus at the TDT point (see Section 7.6 and Fig. 7.14) is analyzed. This torus is non-smooth and the law (5.19) is valid. Finally, we analyze the SNA in the GOPY-model (2.11, 2.12) presented in Fig. 2.1. If we choose here an observable $y_n = |x_n|$, then y_n is a function of θ_n given by an upper discontinuous curve (this has been proven by Keller [1996]). The spectral characteristics for this observable is shown in Fig. 5.3. Similar to Fig. 5.2, the number of harmonics that exceed a certain level grows as a power law (5.19).

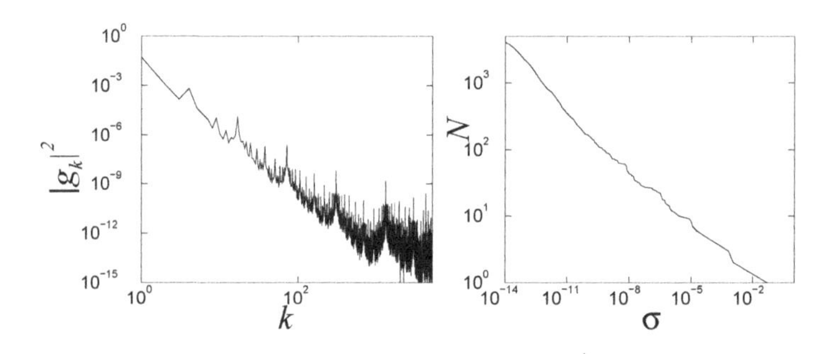

Fig. 5.2 Spectral characteristics of the critical torus shown in Fig. 7.14 for the observable x_n. Left panel: Fourier coefficients; right panel: the number of Fourier harmonics $N(\sigma)$ that exceed the level σ.

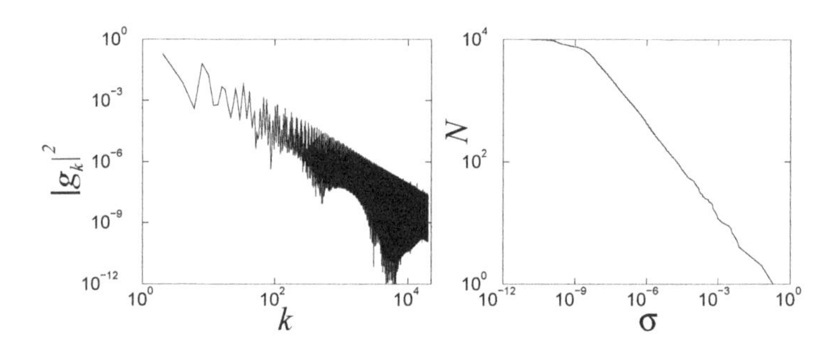

Fig. 5.3 Spectral characteristics of the attractor shown in Fig. 2.1a for the observable $y_n = |x_n|$. Left panel: Fourier coefficients; right panel: the number of Fourier harmonics $N(\sigma)$ that exceed the level σ.

5.2.3 *Singular continuous spectrum in an SNA*

In many situations an SNA cannot be considered as a single-valued function (5.10), and in this case the characterization based on the calculation of the discrete spectral components as outlined in Section 5.2.2 is not valid. At the moment there is no general theory of correlations and spectra of strange nonchaotic attractors. This is mainly due to the reason mentioned above: in the spectrum generally a discrete quasiperiodic component is present corresponding to the forcing, and it is hard to separate it to study the rest of the spectrum. There is, however, a system, where this discrete component vanishes due to a special symmetry of the observations; this system is the GOPY-map

$$x_{n+1} = 2a\tanh(x_n)\cos(2\pi\theta_n) , \qquad (5.20)$$

$$\theta_{n+1} = \theta_n + \omega . \qquad (5.21)$$

Here the observable x_n does not possess a discrete spectrum. Below we describe the correlation and spectral properties of system (5.20, 5.21) in detail, first numerically and then analytically.

The main point is that the spectrum of (5.20, 5.21) is singular continuous, or fractal. In this way it lies in between the discrete and the non-singular continuous (broad band) one (in order to simplify the presentation we will call the latter spectrum with finite spectral density simply "continuous spectrum"). We will dwell on the corresponding features below.

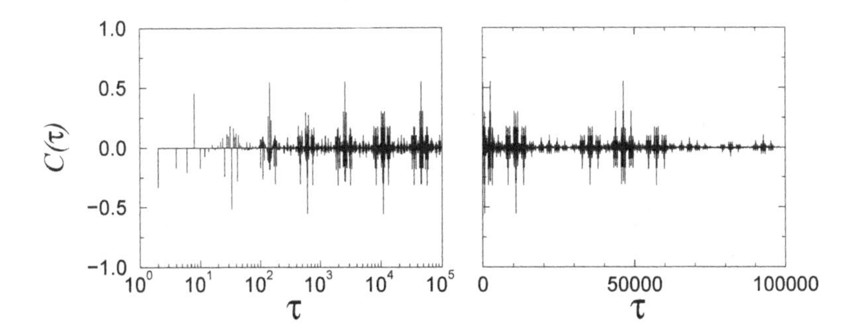

Fig. 5.4 Autocorrelation function of the observable x_n for system (5.20, 5.21), in logarithmic (left panel) and linear (right panel) time scales.

The (normalized) autocorrelation function for the observable x_n is shown in Fig. 5.4. The correlations neither come close to one (like in the case of a purely discrete spectrum) nor decay to zero (like for a purely continuous spectrum). There are finite peaks at a level $\approx \pm 0.55$ which, however, are not periodic in time but occur at larger and larger intervals between them. This is mostly clear in the left panel of Fig. 5.4, where the time axis is logarithmic – here the peaks appear periodically. Remarkably, not only the major peaks, but also the structures around them appear to be nearly periodic, this suggests a high degree of self-similarity in the correlations, see Fig. 5.5.

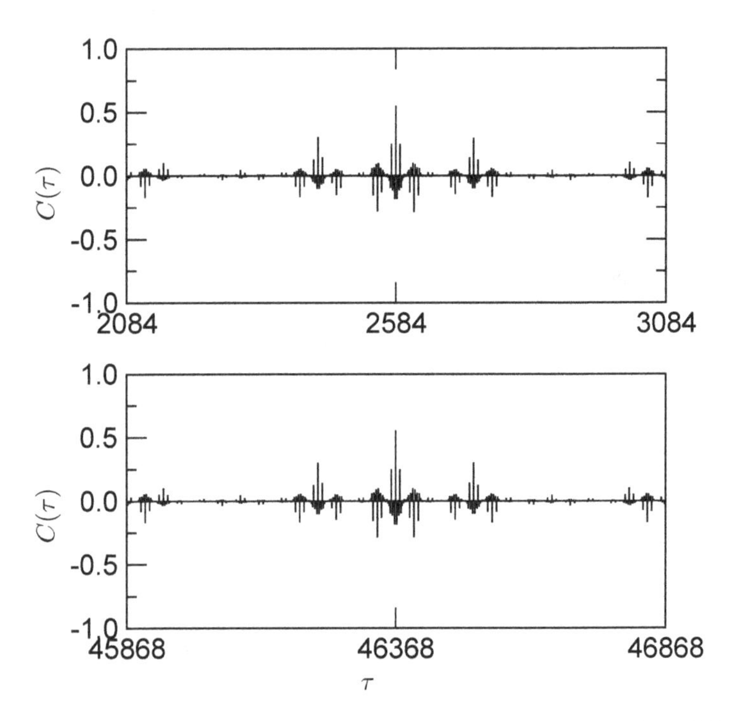

Fig. 5.5 The fine structure of the autocorrelation function near the peaks at Fibonacci numbers $\tau = 2584$ and $\tau = 46368$ shows self-similarity.

The spectrum corresponding to the correlation function Fig. 5.4 is a fractal. Because it is difficult to draw a fractal function, the usual way to represent it is to look at a sequence of approximations to it. For the power spectrum such approximations are naturally given by the finite Fourier transforms (5.13), or, alternatively, one can use the finite Fourier transforms of the correlation function

$$S(\Omega, T) = \sum_{k=-T}^{T} \left(1 - \frac{k}{T}\right) \cos(2\pi\Omega k) C(k) \; . \tag{5.22}$$

These transforms for different time intervals T are shown in Fig. 5.6. One can see that the number of peaks and their heights grow with T, so that in the limit $T \to \infty$ one can imagine a function with singularities at every point (we shall discuss the nature of these singularities below).

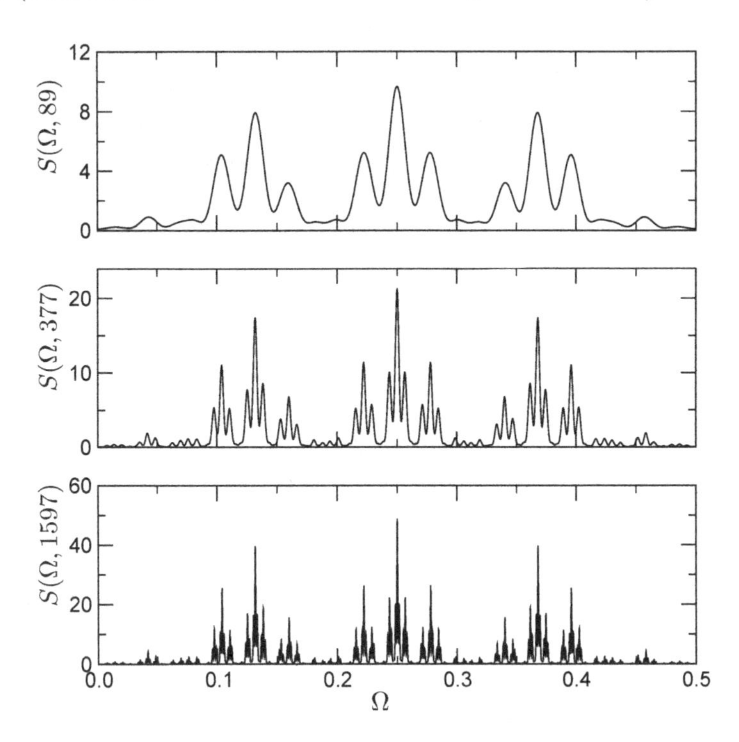

Fig. 5.6 Approximations to the power spectrum of the observable x_n in (5.20, 5.21) according to formula (5.22) for three different values of T. With increasing T the spectrum gets more and more fine: the peaks grow with T (notice different scales of the vertical axes) and their number increases.

A natural characterization of the fractal spectrum is via its dimension characteristics. A simple way to calculate the correlation dimension D_2 of the spectrum is given by the Ketzmerick formula [Ketzmerick et al. 1992], which relates this dimension with the behavior of the integrated correlation function defined as

$$C_{int}(T) = \frac{1}{T}\sum_{\tau=0}^{T} C^2(\tau) . \tag{5.23}$$

The dimension D_2 is related to C_{int} according to the asymptotic (for large T) relation

$$C_{int}(T) \sim T^{-D_2} . \tag{5.24}$$

One can easily see that for a continuous spectrum the sum in (5.23) converges, so that $C_{int}(T) \sim T^{-1}$ and $D_2 = 1$, while for a discrete spectrum the correlations do not decay, the sum grows $\sim T$ and $D_2 = 0$. For a singular continuous spectrum one expects $0 < D_2 < 1$. The integrated correlation function for the SNA (5.20, 5.21) is shown in Fig. 5.7, it yields $D_2 \approx 0.7$.

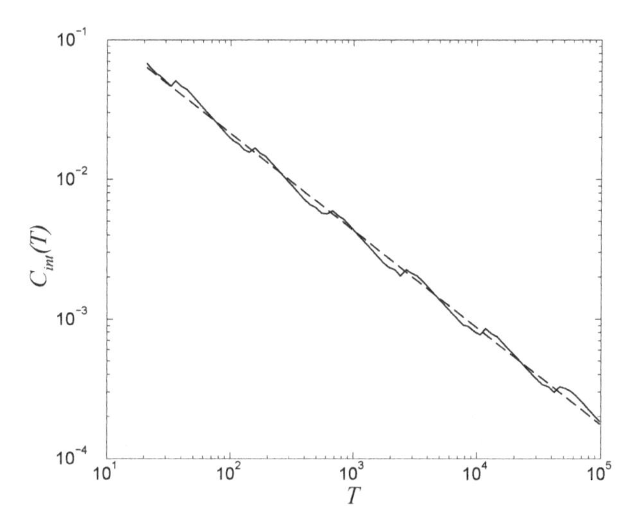

Fig. 5.7 The integrated correlation function computed from Fig. 5.4. Notice the logarithmic periodicity due to the logarithmic–periodic structure of the peaks in Fig. 5.4. The power-law fit (dashed line) yields $D_2 \approx 0.7$.

Now we would like to demonstrate that the peaks in the spectrum Fig. 5.6 are not delta-peaks. To this end we calculate the sums (5.13) for several frequencies corresponding to the peaks in the spectrum. The resulting curves are shown in Figs. 5.8, 5.9. One can see that the spectral sum $s(\Omega, T)$ (5.13) as a function of time T is a self-similar walk on the complex plane. This walk should be compared with the cases of continuous and discrete spectra. For a continuous spectrum the corresponding curve is a random walk whose distance from the initial point grows as $|s(\Omega, T)| \sim \sqrt{T}$, thus after normalizing like in (5.14) one gets a finite spectral density at frequency Ω. For a discrete spectral component at frequency Ω one observes a directed motion on the complex plane, the velocity of this motion gives the amplitude of the delta-peak in the spectrum. We see that in the cases presented in Figs. 5.8, 5.9 the motion is in between these two cases: one observes a self-similar curve, whose deviation from the origin grows slower than $\sim T$ but faster than $\sim \sqrt{T}$. We show the squared distance from the origin $|s(\Omega, T)|^2$ as a function of time in Fig. 5.10. The curves are logarithmically periodic due to self-similarity, the slopes give the relative strengths of the peaks in the power spectrum.

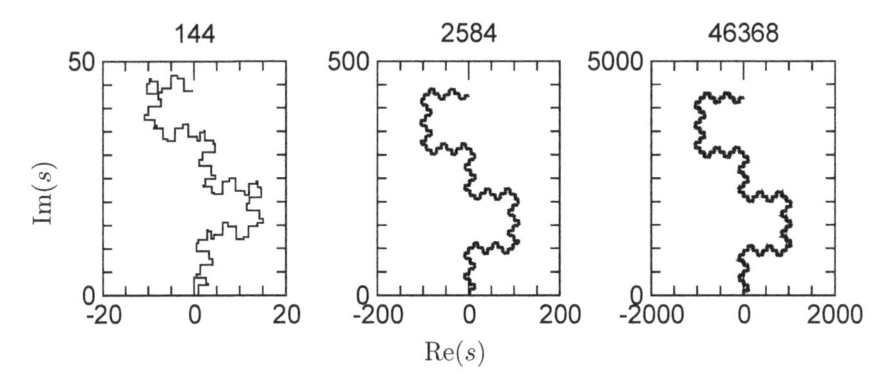

Fig. 5.8 Spectral sum $s(\Omega, T)$ for $\Omega = \frac{1}{4}$ defined according to (5.13) on different time intervals $T = 144$, $T = 2584$, and $T = 46368$. One can see that the "random walk" on the plane of real and imaginary parts of s is self-similar. Notice that the scales on the panels differ by factor 10 while the time intervals differ by factor ≈ 18.

5.2.4 *Theoretical description of the singular continuous spectrum*

In this section we will explain the numerical results for the correlations and spectra of SNA theoretically. The theory is rather involved, therefore we will not go into all detail, but rather present a sketch of the main ideas.

The first step is to choose a suitable observable for the system (5.20, 5.21). Instead of looking on the observable x_n, we will consider

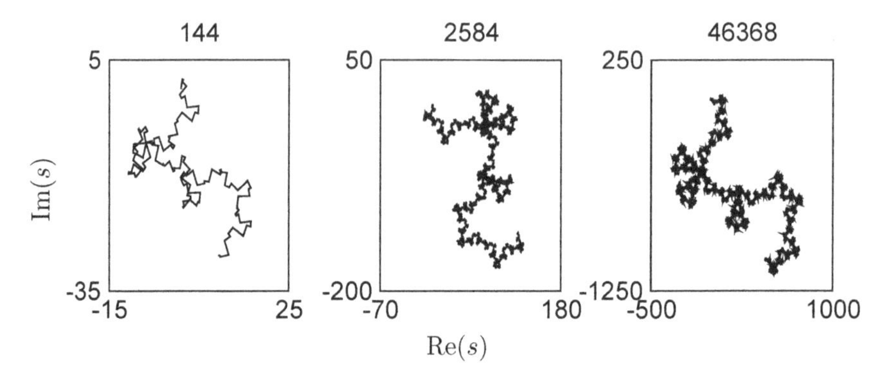

Fig. 5.9 The same as in Fig. 5.8, but for frequency $\Omega = \frac{5+\sqrt{5}}{10}$.

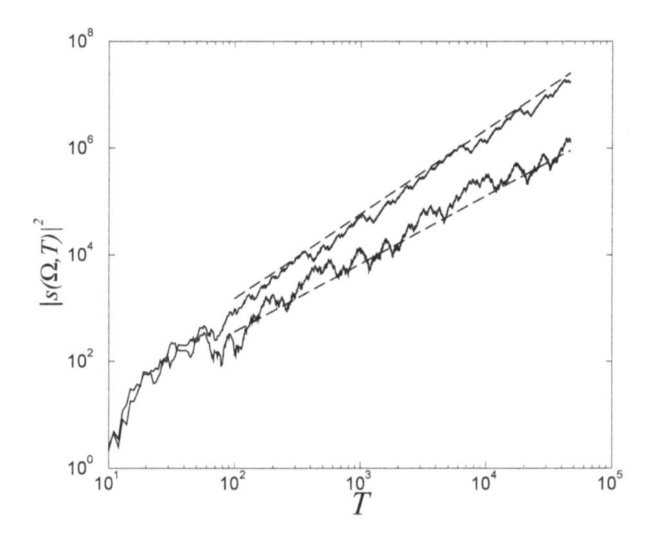

Fig. 5.10 Growth of spectral sums (5.13) for the two frequencies presented in Fig. 5.8 ($\Omega = \frac{1}{4}$, upper curve) and Fig. 5.9 ($\Omega = \frac{5+\sqrt{5}}{10}$, lower curve). The dashed lines show the slopes of the power-law growth: $|s(\Omega,T)|^2 \sim T^{1.588}$ for $\Omega = \frac{1}{4}$ and $|s(\Omega,T)|^2 \sim T^{1.277}$ for $\Omega = \frac{5+\sqrt{5}}{10}$.

$y_n = -\text{sign}(x_n)$. This new observable is more simple as it takes values 1 and -1 only. Numerical simulations show that the main quantitative properties of the correlations and spectra for y_n are the same as for x_n. For this new observable the time evolution reads

$$y_{n+1} = y_n \Phi(\theta_n) \,, \tag{5.25}$$

$$\theta_{n+1} = \theta_n + \omega \,, \tag{5.26}$$

where the function Φ is

$$\Phi(\theta) = \begin{cases} -1 & \text{if } 0 \le \theta < \frac{1}{2} \,, \\ \;\;1 & \text{if } \frac{1}{2} \le \theta < 1 \,. \end{cases} \tag{5.27}$$

Next we use the self-similar properties of the irrational $\omega = \frac{\sqrt{5}-1}{2}$, using its approximations via Fibonacci numbers. It is convenient first to consider the sequence $z_n = \Phi(\theta_n)$ starting from $\theta_0 = 0$. Note that the observable z_n is a quasiperiodic function of time (cf. (5.10)), but the observable y_n which satisfies

$$y_{n+1} = y_n z_n = y_0 \prod_{k=0}^{n} z_k \tag{5.28}$$

is not quasiperiodic.

The sequence z_n can be produced from the following inflation rule (this rule has been derived in [Aubry et al. 1988]):

$$\begin{aligned} A &\to T(A) = CAC \,, \\ B &\to T(B) = CACCA \,, \\ C &\to T(C) = CACBA \,. \end{aligned} \tag{5.29}$$

Here three symbols are present, but eventually one should set $A = B = -1$, $C = 1$. Starting from the initial symbol $z_1 = C$, the inflation rule (5.29) generates the sequence z_n. There is also another representation of self-similarity in the sequence z_n, namely via a concatenation rule. This means that having a part of the sequence z_n, one can continue its construction by adding some already present pieces. If we denote with Z_N the sequence z_n of length F_N equal to some Fibonacci number, then according to [Feudel et al. 1996b]

$$Z_{n+1} = \begin{cases} Z_n Z_{n-3} Z_{n-4} Z_{n-3} = Z_{n-1} Z'_n & \text{if } n = 3k+1, \; k \ge 2 \,, \\ Z_n Z_{n-1} & \text{otherwise} \;\;, \end{cases} \tag{5.30}$$

where the prime means that the last symbol in the string is inverted. This representation is the basis for the self-similarity in the power spectrum. We do not write out the rather complex concatenation rule that allows one to construct the sequence y_n, it can be found in [Feudel et al. 1996b].

The main idea now is that the spectrum of the sequence can be also constructed iteratively, as the Fourier transform of a new portion of the sequence can be represented via transforms of already present portions. For the sequence y_n this leads to the following results.

The non-normalized Fourier transform of the sequence $Y_N = [y_1, ..., y_{F_N}]$ is defined as (cf. (5.13))

$$s_N(\Omega) = \sum_{l=1}^{F_N} y_k e^{i2\pi l\Omega} . \tag{5.31}$$

The concatenation rule for y_n allows for obtaining a recursive relation for $s_N(\Omega)$. Here one should also properly take into account the phases, because the Fourier transform of the portion shifted by F_k comes with a phase factor of type $e^{i2\pi F_k \Omega}$. The resulting transformation of the Fourier spectrum is rather cumbersome:

$$\begin{aligned}
s_{6k+3} &= s_{6k+2} + e^{i\phi_{6k+2}} s_{6k+1} , \\
s_{6k+4} &= s_{6k+3} - e^{i\phi_{6k+3}} s_{6k+2} , \\
s_{6k+5} &= s_{6k+3} - e^{i\phi_{6k+3}} s_{6k+4} - 2e^{i\phi_{6k+5}} , \\
s_{6k+6} &= s_{6k+5} - e^{i\phi_{6k+5}} s_{6k+4} , \\
s_{6k+7} &= s_{6k+6} + e^{i\phi_{6k+6}} s_{6k+5} , \\
s_{6k+8} &= s_{6k+6} + e^{i\phi_{6k+6}} s_{6k+7} + 2e^{i\phi_{6k+8}} ,
\end{aligned} \tag{5.32}$$

where, for the sake of brevity, we omit the Ω-dependence, and

$$\phi_n \equiv 2\pi F_n \Omega$$

satisfies

$$\phi_n = \phi_{n-1} + \phi_{n-2} \pmod{2\pi} . \tag{5.33}$$

The initial conditions for (5.32, 5.33) are given as

$$s_0 = 0 \quad , \quad s_1 = -e^{i2\pi\Omega} , \qquad \phi_0 = 0 \quad , \quad \phi_1 = 2\pi\Omega . \tag{5.34}$$

The asymptotic behavior of s_n gives the growth rate of the spectral component at a given frequency Ω. It depends crucially on the evolution of ϕ. Mapping (5.33) is Arnold's cat map: it has chaotic trajectories and an everywhere dense set of periodic orbits. To which orbit the sequence $\{\phi\}$ converges, depends on the initial conditions (5.34), i.e. on the frequency Ω. If ϕ is eventually attracted to a periodic orbit having period m (i.e., if the initial point lies on the stable manifold of this orbit), then system

(5.32) can be seen as a periodically driven linear map. This is the case for the frequencies chosen for Fig. 5.10. Accordingly, when the evolution is monitored every Kth iteration, where K is the least common multiple of m and 6, we find that it is described by a time-independent[2] linear map acting on a 2-dimensional space (in fact, the updating of s_{n+1} always requires the knowledge of s_n and s_{n-1} only).

It is easily seen that the determinant of the homogeneous part of the transformation (5.32) has modulus one in each of the six steps. Accordingly, volumes are conserved and the moduli of the two eigenvalues of each periodic orbit are inverse to each other. Therefore, the growth rate Λ of s_n, i.e. the largest Lyapunov exponent of map (5.32), cannot be smaller than 0. Correspondingly, the normalized spectral component $|s(\Omega, T)|^2$ grows with time T as $|s(\Omega, T)|^2 \sim T^{\gamma}$, where

$$\gamma(\Omega) = 2\frac{\Lambda(\Omega)}{\log(1/\omega)} . \tag{5.35}$$

(Here we use $F_N \sim \omega^{-N}$.) This power-law behavior is precisely the growth rate that has been illustrated in Fig. 5.10. The values of the exponent $\gamma = 1.58796$ and $\gamma = 1.27644$ calculated in [Feudel et al. 1996b] agree perfectly with the fits in Fig. 5.10.

5.3 Bibliographic notes

Ding, Grebogi, and Ott [1989a] first discussed and calculated numerically different dimensions of SNA. Later Datta et al. [2003] applied a multifractal formalism to characterize fractal properties of SNAs. Keller [1996] has proven that in the GOPY model the graph of $|x|$ vs. θ is an upper discontinuous curve.

Ding et al. [1989b] calculated the power spectrum for a quasiperiodically forced circle map. The singular continuous spectrum was found in [Pikovsky and Feudel 1994], in [Feudel et al. 1996b] a renormalization analysis of correlations and the spectrum has been performed, later Mestel and Osbaldestin [2000] considered this problem more rigorously. Bezhaeva and Oseledets [1996] proved that the GOPY model possesses a singular continuous spectrum in some range of parameters.

[2] Here, we mean the renormalization time.

Chapter 6

Bifurcations in quasiperiodically forced systems and transitions to SNA

Nonlinear dynamical systems usually exhibit different long-term behaviors in different regions of relevant system parameters. Such long-term behaviors can be stationary points, periodic orbits, quasiperiodic motion or chaotic attractors. As discussed above, an additional class of attractors, namely strange nonchaotic attractors, occurs typically in quasiperiodically forced systems. Thus the variety of possible long-term behaviors in quasiperiodically forced systems is larger than in other classes of nonlinear dynamical systems because of the additional existence of SNAs. When system's parameters are varied, transitions between different long-term behaviors can occur. Such transitions, called bifurcations, happen usually when critical threshold values for particular parameters are crossed. In this chapter we discuss mechanisms of various bifurcations occurring in quasiperiodically forced systems. Our particular focus will be on such bifurcations which lead to the emergence of SNAs. In general, SNA are found in tiny parameter intervals close to the transition to chaos. Therefore we investigate the three universal routes to chaos under the influence of quasiperiodic forcing. We study the period doubling route, the route via quasiperiodicity and the intermittency route as the major transitions to chaos and relate them to the emergence of SNA. However, there are also systems like the GOPY-map (Section 2.2.1) where an SNA exists for large regions in parameter space without any transition to chaos.

To demonstrate different kinds of bifurcations we use various paradigmatic models introduced in Chapter 2, such as the quasiperiodically forced logistic, GOPY, circle and intermittency maps, where most of the transitions can be observed. Of course, there are many other model systems where these bifurcations can occur, but we restrict ourselves to a very few systems to give a comprehensive overview, and mention other systems studied by other authors in the bibliographic notes.

The simplest attractor in a quasiperiodically forced system corresponds to a quasiperiodic motion on a torus in the case of flows and invariant curves in the case of maps, while strange nonchaotic and chaotic attractors are more complex. There are several mechanisms leading to the emergence of SNA, but they share the common property that SNAs are always formed out of quasiperiodic attractors. The structure and the statistical properties of SNA tell us already something about the possible mechanisms for the transition to SNA: Sturman and Stark [1999] and Glendinning [1998] have demonstrated that in some cases an unstable invariant set is embedded in an SNA. This corresponds to the finding that the distribution of finite time Lyapunov exponents has a positive tail, i.e. SNA do not exhibit uniform contraction but are always characterized by embedded regions of expansion (cf. Chapter 4). Thus the emergence of SNA is in many cases connected with the collision of a stable invariant set with an unstable one, where the latter becomes embedded in the SNA beyond the transition. The only exception from this thumb rule so far is the appearance of SNA due to the fractalization of the torus [Kaneko 1986; Nishikawa and Kaneko 1996].

This chapter introduces different known transitions to SNA and discusses characteristic features of these transitions. Furthermore, other important bifurcations in quasiperiodically forced systems are explained. For the sake of simplicity we choose only maps to describe the bifurcations, but most of them can be obtained in flows as well. Therefore we assign in this chapter always the term torus to a quasiperiodic motion, also when discussing invariant curves in noninvertible maps.

6.1 Smooth and non-smooth bifurcations

Quasiperiodic forcing adds two different frequencies to the motion in the system under consideration. Due to the forcing, periodic orbits of the unforced system turn into invariant curves for maps and tori T^2 for flows, as already explained in Chapter 2. Periodic orbits with period 2^m are transformed into invariant curves which consist of 2^m branches or tori with multiplicity 2^m. Keeping this in mind, we expect to find all major known bifurcations which occur for fixed points or periodic orbits in unforced systems. Such bifurcations would be saddle-node, transcritical and pitchfork bifurcations as well as period doublings and the Neimark-Sacker bifurcation. Indeed, we can find all these bifurcations when the forcing amplitude is small. However, in a quasiperiodically forced system these bifurcations do not happen to fixed points or periodic orbits, but to tori (for a mathematical treatment of bifurcations in quasiperiodically forced systems see [Broer et al. 1990]).

Saddle-node, pitchfork and transcritical bifurcation: A saddle-node bifurcation (tangent bifurcation) which involves a stable and an unstable fixed point in the unforced system corresponds to a merging of a stable and an unstable torus in the quasiperiodically forced system. Merging means here that the two tori come close to each other and merge smoothly in each point. The same happens in transcritical or pitchfork bifurcations.

Period doubling: A period doubling in the unforced system turns into a torus doubling in the forced system. Prior to the bifurcation we have a torus with multiplicity 2^m and beyond the bifurcation a torus with multiplicity 2^{m+1} is obtained. We denote this transition by $T2^m \to T2^{m+1}$.

Neimark-Sacker bifurcation: A bit more complicated is the Neimark-Sacker (secondary Hopf) bifurcation where in the unforced system a periodic orbit loses its stability and a stable quasiperiodic motion on T^2 with two incommensurate frequencies arises. The equivalent bifurcation in a quasiperiodically forced system is the loss of stability of a two-frequency torus T^2 leading to the emergence of a stable motion on a three-frequency torus T^3. On such tori T^3 phase-lockings can occur, where the three-frequency motion turns into a two-frequency one topologically confined to the torus T^3. Such phase-lockings correspond to a saddle-node bifurcation of two smooth tori T^2.

All aforementioned bifurcations we will call *smooth* bifurcations since they happen when two or more tori merge smoothly in each point. A detailed description of these smooth bifurcations is given in the next Sections.

However, large amplitudes of quasiperiodic forcing lead to a wrinkling of the tori and thus to *non-smooth* bifurcations in the sense that the tori do not merge in each point but only collide in a dense set of points. These bifurcations, which often give rise to the formation of strange nonchaotic attractors, are specific to quasiperiodically forced systems and will be discussed in detail in the next Sections.

6.2 Bifurcations in the quasiperiodically forced logistic map

The simplest nonlinear map studied in chaos theory is the one-dimensional logistic map. It is one of the most extensively studied maps in nonlinear dynamics, since it exhibits the universal transition to chaos via an infinite period doubling cascade. Because strange nonchaotic attractors are expected to appear in the neighborhood of the transition to chaos, the quasiperiodically forced logistic map is the best paradigmatic example to examine the influence of quasiperiodic forcing on the bifurcations known in the unforced logistic map, in particular on the period doubling route to chaos. Moreover, several other mechanisms for the formation of SNAs occur in this simple map.

The quasiperiodically forced logistic map is used in two different forms which are equivalent, since they can be mapped onto each other via a coordinate transformation (cf. Section 2.2.4). On one hand we use the quadratic map in the following form:

$$x_{n+1} = a - x_n^2 + \varepsilon \cos 2\pi\theta_n \,, \tag{6.1}$$

$$\theta_{n+1} = \theta_n + \omega \pmod 1 \,, \tag{6.2}$$

on the other hand we also use the logistic map and replace (6.1) by

$$x_{n+1} = r x_n (1 - x_n) + \varepsilon \cos 2\pi\theta_n . \tag{6.3}$$

The parameter a respective r measures the strength of the nonlinearity, while ε denotes the amplitude of the forcing. The forcing frequency is fixed at the inverse of the golden mean $\omega = (\sqrt{5} - 1)/2$. Fig. 6.1 shows the bifurcation diagram in the parameter space spanned by a and ε for Eqs. (6.1, 6.2). We can distinguish various regions which correspond to different long-term behaviors. The simplest attractors are associated with quasiperiodic motions with two incommensurate frequencies. In Fig. 6.1 we can identify quasiperiodic motion on a single torus denoted by $T1$, motion on a doubled torus $T2$ and on a quadrupled torus $T4$. Furthermore, there are regions where we find strange nonchaotic attractors, marked by SNA and chaotic motion indicated by Ch. The white region in parameter space represents all those parameter values where the trajectories diverge to infinity. All regions corresponding to different long-term behavior are separated by boundaries which are related to bifurcations leading to transitions from one long-term behavior to another. All the transitions explained and illustrated in the following Subsections are indicated by arrows in Fig. 6.1.

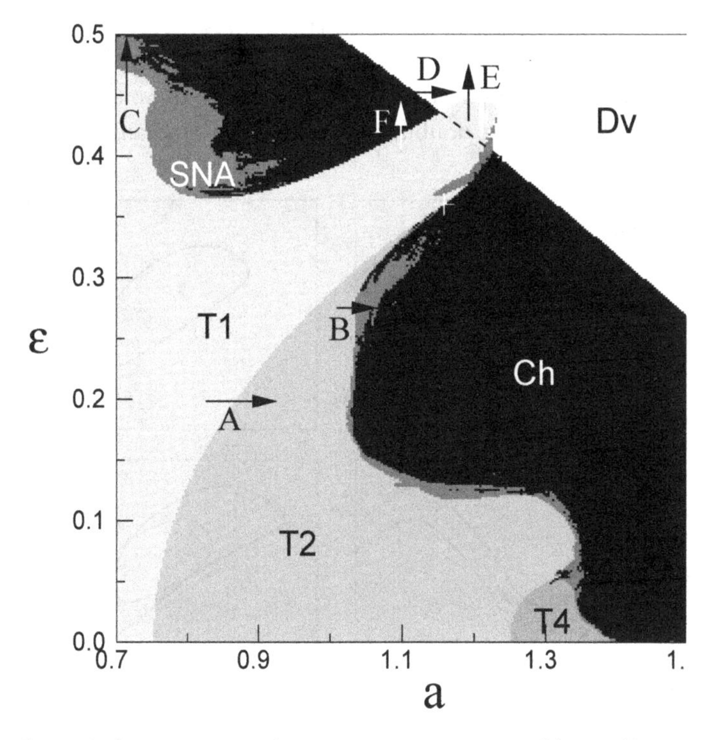

Fig. 6.1 Dynamical transitions in the parameter space spanned by nonlinearity a and forcing amplitude ε. $T1$: single torus, $T2$ doubled torus, $T4$: quadrupled torus, Ch: chaos, Dv: divergence of the trajectory. Possible transitions which are explained in this Chapter: A: torus doubling, B: non-smooth tori collision, C: fractalization of the torus, D: boundary crisis of the chaotic attractor, E: boundary crisis of the torus, F: intermittency transition (resembling an interior crisis). The dashed line indicates the basin boundary bifurcation.

6.2.1 *Torus doubling*

A torus doubling in quasiperiodically forced systems is the smooth equivalent of a period doubling in an unforced system. Crossing the bifurcation line along route A in Fig. 6.1, the single torus loses its stability and a stable double torus emerges. An illustration of this scenario is shown in Fig. 6.2 in two different projections. While in the (θ, x) state space (Fig. 6.2, left column) the single torus corresponds to one branch, the doubled torus consists of two branches, which have to be understood using the mod 1 operation in θ. In the (x_{n-1}, x_n) projection (Fig. 6.2, right column) we obtain closed invariant curves. It has been shown by Heagy and Hammel [1994] that the torus doublings are delayed, i.e. they are shifted to higher values of the nonlinearity parameter a compared to the unforced case. Therefore, the curves of torus doublings are bended in the (a, ε) parameter space Fig. 6.1. The larger the forcing amplitude ε, the larger needs to be the value of nonlinearity a to find the torus doubling. Qualitatively the same picture results

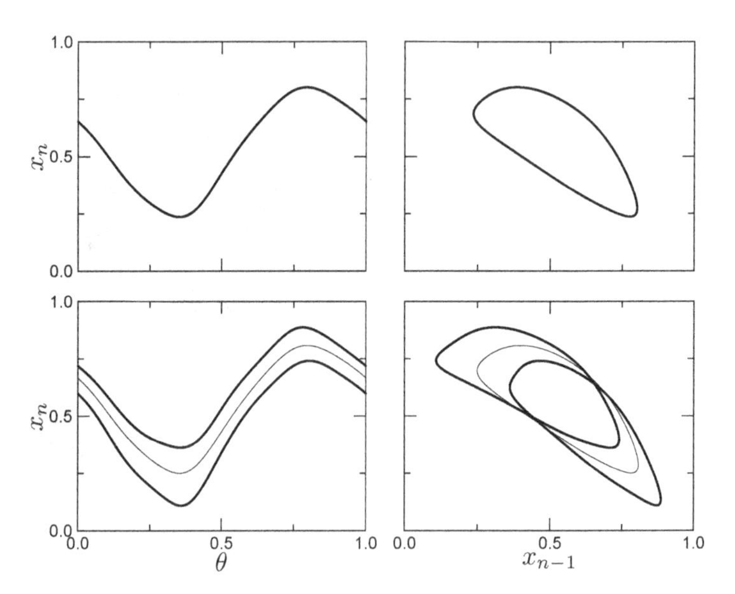

Fig. 6.2 Smooth torus doubling bifurcation in the quadratic map ($\varepsilon = 0.2$) in the (θ, x) plane (left) and the (x_{n-1}, x_n) plane (right): single torus prior to the bifurcation (top panels) for $a = 0.86$, beyond the bifurcation (bottom panels) for $a = 0.88$. The stable torus is shown with bold dots, the unstable one with normal dots.

for all subsequent torus doublings. Fig. 6.1 shows also the bifurcation from a doubled torus to the quadrupled torus, where the torus with 2 branches loses stability and a stable torus with 4 branches appears [Sosnovtseva et al. 1996].

In contrast to the infinite period doubling cascade in unforced systems, the torus doubling cascade in quasiperiodically forced systems is always *finite*. The quasiperiodic forcing leads to a suppression of torus doublings. The pruning of the torus doubling cascade is related to the emergence of SNA due to another bifurcation which is described in detail in the next subsection. How many torus doublings can be observed, depends on the strength of the forcing. With increasing forcing amplitude fewer and fewer torus doubling bifurcations occur and the torus doubling cascade ends with tori of lower and lower multiplicity. In Fig. 6.1 this becomes apparent when looking at the two lines of torus doublings. They both end at a certain value for the forcing amplitude. This terminal point of the torus doubling (TDT point), which will be discussed in detail in Chapter 7, indicates, that for higher forcing amplitudes this torus doubling does not occur anymore, since the torus has turned already into another kind of attractor which is not capable of further doubling. The higher the multiplicity of the torus, the lower is the forcing amplitude at which the TDT point appears. This becomes obvious comparing the torus doubling lines $T1 \to T2$ and $T2 \to T4$ in Fig. 6.1. Thus the torus doublings with high multiplicity are pruned first, yielding a finite cascade of torus doublings.

6.2.2 *Non-smooth tori collision beyond period-doubling*

Besides the appearance of a terminal point of torus doublings, the suppression of further doublings is related to the emergence of SNA due to a mechanism which has been reported by Heagy and Hammel [1994]. This mechanism of creation of an SNA occurs in the whole class of quasiperiodically forced systems possessing a period doubling cascade without forcing. In the quasiperiodically forced quadratic map this transition occurs along the route B indicated in Fig. 6.1. Without loss of generality we focus here on the first torus doubling because for all higher ones the mechanism is the same but occurs for smaller forcing amplitudes ε.

Suppose we consider the system just beyond the torus doubling bifurcation (cf. Section 6.2.1). Then we observe a stable doubled torus $T2$ consisting of two branches in the (θ, x) space and an unstable torus $T1$ with only one branch. The unstable torus is located in the middle between the two stable branches. With increasing nonlinearity a the branches of the stable $T2$ become more and more wrinkled, while the unstable torus $T1$ remains smooth (Fig. 6.3a). At some critical nonlinearity a_c the two stable branches touch its unstable parent in a dense set of θ values. As a result of this non-smooth collision between the stable torus $T2$ and the unstable torus $T1$ a strange nonchaotic attractor is born (Fig. 6.3b). The unstable torus $T1$ becomes embedded in the SNA leading to local expansion of nearby trajectories. Nevertheless, beyond the collision of the two invariant sets the Lyapunov exponent remains negative (see Fig. 6.4. Only a further increase of the nonlinearity leads to a positive Lyapunov exponent and thus to chaotic motion which is demonstrated in Fig. 6.3c.

The same scenario happens for all torus-doubling bifurcations. In general, we observe transitions $T2^m \to$SNA2^{m-1} depending on the nonlinearity

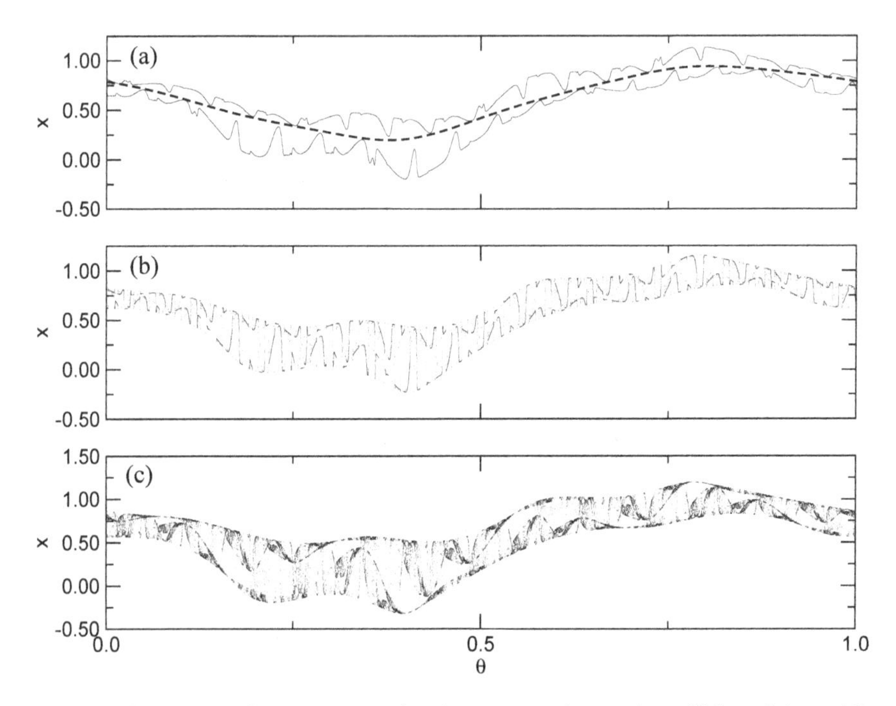

Fig. 6.3 Emergence of a strange nonchaotic attractor due to the collision of the stable doubled torus with its unstable parent in the quasiperiodically forced quadratic map for $\varepsilon = 0.275$: (a): $a = 1.03$ just prior to the transition, the unstable torus is depicted with the dashed line; (b): $a = 1.04$ beyond the transition in the regime of SNA, where now the attractor contains both former branches of the torus as well as the unstable torus; (c): $a = 1.07$ beyond the transition to chaos. The values of parameter a correspond to those marked in Fig. 6.4.

or the forcing amplitude. This means that at the transition, a torus with 2^m branches turns into an SNA with 2^{m-1} branches. The higher the multiplicity of the torus, the smaller is the forcing amplitude needed to cause the transition to a strange nonchaotic attractor. After the appearance of SNA no further doublings are possible, i.e. the torus-doubling cascade is pruned. As a consequence, the torus doubling cascade becomes shorter and shorter with increasing forcing. As already mentioned above, the bifurcation curve of each torus doubling in the (a, ε) plane ends in a specific terminal point (TDT point) which will be discussed in detail using a renormalization group approach in Chapter 7.

It has to be noted that Heagy and Hammel [1994] studied the quasiperiodically forced logistic map in a slightly different form. Instead of an additive forcing term they use a multiplicative one. The disadvantage of this

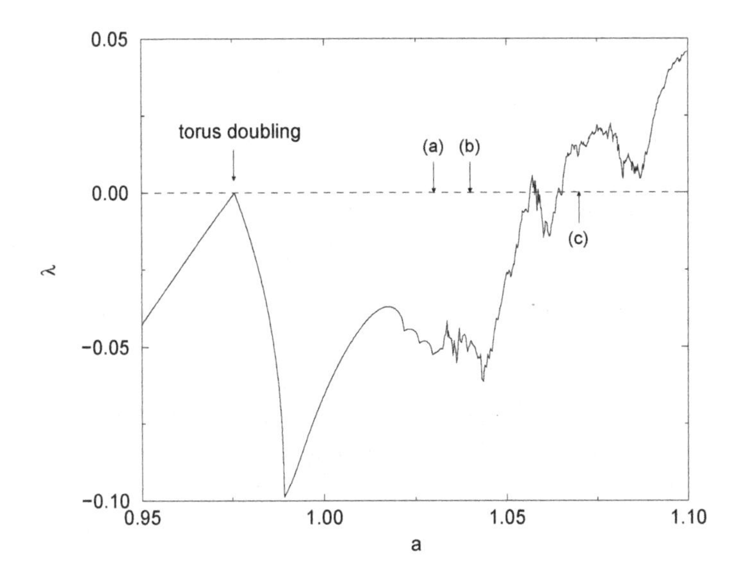

Fig. 6.4 Lyapunov exponents depending on nonlinearity parameter a indicating the parameter values of the phase portraits shown in Fig. 6.3: doubled torus $T2$ (a), SNA (b) and chaos (c).

form is that the two parameters a respective r for the nonlinearity and ε for the forcing amplitude are not independent in their action since they are multiplicatively connected. Since all their findings occur also in the usual quasiperiodically forced quadratic map Eqs. (6.1, 6.2), we have employed that one to demonstrate the appearance of strange nonchaotic attractors. The emergence of SNA due to the collision of a doubled torus with its unstable parent has also been studied in the Hénon and the ring map [Sosnovtseva et al. 1996], where also torus doublings with higher multiplicity are studied.

6.2.3 *Fractalization of the torus*

The fractalization of the torus is the most common but also the most intriguing of the possible transitions to SNA. Fractalization means that the torus becomes more and more wrinkled as forcing increases, until it breaks up to form a strange set without becoming chaotic. This mechanism of the formation of strange nonchaotic attractors has been described by Kaneko already in 1984 even though he did not term the emerging new invariant set a strange nonchaotic attractor [Kaneko 1984b, 1986]. The process of fractalization appears in many models as a gradual change in the structure of the attractor which is difficult to relate to a precise bifurcation point, where a sudden change in the dynamics occurs due to the crossing of a well-defined critical threshold. The fractalization transition to SNA is also exceptional from another point of view: In contrast to all other known mechanisms for the emergence of strange nonchaotic attractors there is no obvious unstable invariant set involved in this transition. Thus it is still an open question what exactly causes the strange structure in this case. However, in some particular cases, namely for forced noninvertible maps, one can define a critical threshold for the fractalization transition. These critical cases which are denoted by the torus collision terminal point (TCT) and the torus fractalization point (TF) occurring on the bifurcation lines of smooth $\leftrightarrow$ non-smooth saddle-node bifurcations of tori (see Section 6.3.2) and the intermittency transition line (see Section 6.5) respectively are discussed in detail in Chapter 7.

The easiest way to analyze this transition is the computation of the sensitivity exponent explained in Chapter 4. Using the method of computing the maximum derivative along the attractor we can illustrate the fractalization of the torus for the quasiperiodically forced logistic map (6.3, 6.2). For rather large forcing already the first torus doubling is suppressed due to the formation of a strange nonchaotic attractor (cf. arrow C in Fig. 6.1). This situation is presented in Fig. 6.5a,b. Since the transition to SNA is a gradual change of wrinkling, it is sometimes difficult to decide when the transition occurs since a rather wrinkled invariant curve looks quite similar to an SNA. Nevertheless, calculating the phase sensitivity exponent (see Chapter 4 for details) gives a strict criterion for the transition. Fig. 6.6 shows the time evolution of the maximum derivative computed prior to and beyond the bifurcation. Since beyond the transition to SNA no saturation of the derivative occurs, we conclude that the derivative does not exist and thus, the invariant set shown in Fig. 6.5b is strange. The fractalization transition is usually close to the transition to chaos. Thus a chaotic attractor occurs for further increase of the forcing amplitude (Fig. 6.5c). It is important to note that chaotic attractors are also characterized by a diverging derivative. Thus the two attractors shown in Fig. 6.5b and Fig. 6.5c have to be distinguished in their dynamics by computing the Lyapunov exponents.

Other measures to check for a strange structure during the fractalization transition have been introduced in [Kaneko 1986; Nishikawa and Kaneko 1996]. One of those methods is based on the idea that a smooth curve

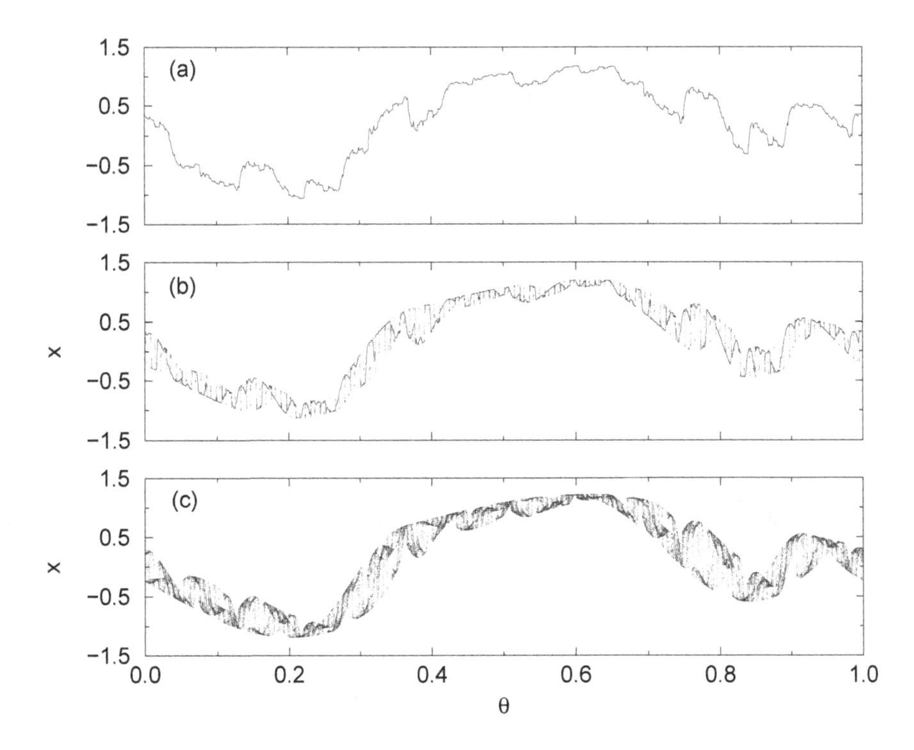

Fig. 6.5 Emergence of an SNA due to the fractalization of a torus in the quasiperiodically forced quadratic map ($a = 0.71$): (a) smooth torus prior to the transition $\varepsilon = 0.47$; (b) SNA beyond the transition $\varepsilon = 0.49$; (c) chaotic attractor $\varepsilon = 0.51$.

has a defined length but a fractal curve does not. For fractals the length depends on the scale used to measure it [Mandelbrot 1977]. As long as the attractor is still an invariant curve, it can be represented as a graph $x = G(\theta)$. This graph has a definite length. The length of the graph can be numerically estimated by a piecewise linear approximation of the graph with a certain number of points. The number of points used defines the accuracy of approximation and, thus, the scale of measurement. With increasing accuracy the length converges in case of a smooth curve while it diverges for a strange nonchaotic attractor (Fig. 6.7).

6.2.4 *Interior crisis*

An interior crisis is a well-known phenomenon in the dynamics of chaotic systems. It has been discovered first for the logistic map when studying

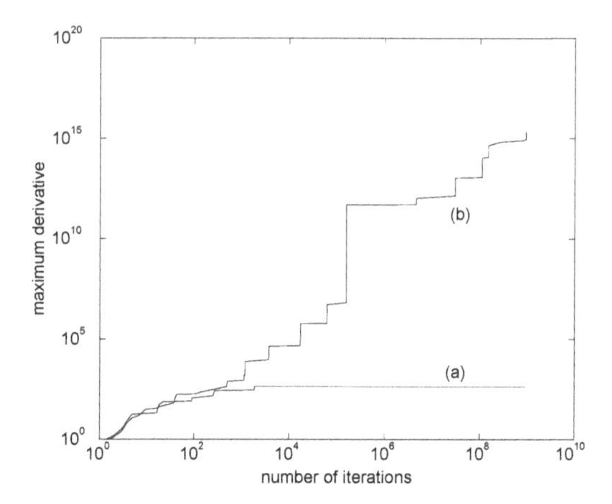

Fig. 6.6 Phase sensitivity for the fractalization transition to SNA depending on the forcing amplitude. The parameters for the curve (a) and (b) correspond to the attractors shown in Fig. 6.5 (a) and (b), respectively.

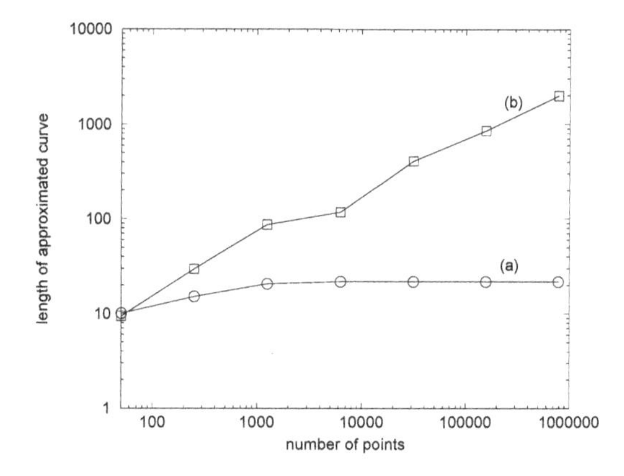

Fig. 6.7 Length of the approximate curve $x = G(\theta)$ vs. accuracy of measurement (number of points used for approximating the invariant curve) for the smooth attractor (a) shown in Fig. 6.5a and the SNA (b) shown in Fig. 6.5b.

the special bifurcations which are responsible for the appearance and disappearance of periodic windows [Grebogi et al. 1982]. In chaotic systems each periodic window is opened due to a saddle-node bifurcation creating a stable and an unstable period-p solution. Varying a particular system parameter, the stable periodic orbit with period p undergoes a period doubling cascade into chaos and finally a chaotic attractor with p pieces appears. In the last bifurcation the window is closed by an *interior crisis* where the unstable period-p orbit (which emerges in the saddle-node bifurcation at the opening of the window) collides with the chaotic attractor consisting of p pieces. As a result a one piece chaotic attractor is formed. The remarkable feature of this chaos–chaos transition is a sudden widening of the attractor as well as a simultaneous reduction of the number of its pieces.

Another kind of interior crisis occurs when a chaotic attractor collides with a chaotic saddle to form a much larger chaotic attractor [Lai et al. 1992]. In this case the number of attractor pieces does not change.

Interior crises have been found likewise in more complicated models [Gallas et al. 1993], as well as in experiments, e.g. [Ditto et al. 1989; Sommerer et al. 1991]. In general, the appearance of an interior crisis corresponds to the collision of a chaotic attractor with an unstable periodic orbit and its stable manifolds or a chaotic saddle. Equivalently, one can study this bifurcation also in terms of a collision of the attractors in a p times iterated system with the boundaries of their respective basins of attraction. Instead of the stable p-periodic orbit beyond the opening of the periodic window one obtains p different fixed points and the separating unstable p-periodic orbit and its stable manifolds make up the boundaries of the basins of attraction of the p fixed points. Moreover, those basin boundaries were found to be very differently structured, sometimes even fractal. Using this transformation, the interior crisis appears then as a simultaneous boundary crisis of all p coexisting attractors. As a consequence the new attractor includes the former attractor pieces and the former chaotic saddles embedded in the basin boundaries.

Let us now discuss how an interior crisis happens in the presence of a quasiperiodic forcing. To study this problem it is necessary to find an extended definition of an interior crisis which grasps the essential properties observed for crises in one-dimensional systems. As interior crisis we denote now a transition with the following properties:

- At the interior crisis there is a sudden widening of the attractor.
- Simultaneously, the number of attractor pieces is reduced.

It should be mentioned that according to this definition an interior crisis does not need to be a transition from chaos to chaos.

In quasiperiodically forced systems 3 different kinds of interior crisis can be found [Witt et al. 1997]: (i) The first kind is the usual one, where first a transition from an SNA to a chaotic attractor both consisting of several pieces occurs. Subsequently they form one large chaotic attractor as a result of an interior crisis. This corresponds to the well known chaos-chaos transition. (ii) The second kind corresponds again to the usual interior crisis in the sense that a sudden enlargement of SNA is observed, i.e. several pieces of SNA built up a large SNA beyond the transition. The only difference to the usual interior crisis is here that the transition involves only nonchaotic motion, since it is an SNA $\rightarrow$ SNA transition. (iii) The third kind, which will be described in detail, is the most unusual one and can be regarded as one mechanism for the appearance of SNA.

To explain the formation of an SNA due to an interior crisis we employ the quasiperiodically forced logistic map (6.2, 6.3) as an example. We focus our study on the dynamics in the period-3 window. The three kinds of interior crisis mentioned above correspond to the 3 arrows in Fig. 6.8, which illustrates a sketch of the dynamical regimes in parameter space for the quasiperiodically forced logistic map. A full study of all three routes can be found in [Witt et al. 1997]. Here we discuss only route (iii) along which an SNA emerges from a quasiperiodic motion on a torus. Fig. 6.9 illustrates this formation of an SNA due to an interior crisis. From a quasiperiodic attractor with 3 branches an SNA with only one branch is formed where the gaps between the former 3 branches are filled with points belonging to the SNA. To explain this mechanism of the appearance of SNA let us look at the structure of the unstable set separating the 3 stable branches of the torus.

For unforced chaotic systems it is known that the chaotic attractor, which exists before the window opens, is transformed into a chaotic saddle that exists throughout the whole window. We assume that in the forced case this chaotic saddle exists as well. But instead of a direct computation of this chaotic saddle we use the concept of fractal basin boundaries to illustrate the existence of this saddle. We start our study at the opening of the window, where two tori with three branches, a stable and an unstable one, emerge in a saddle-node bifurcation. Now we consider only every third iterate of Eqs. (6.2, 6.3), then our 3-times iterated map possesses three different attractors, which are tori with one branch each. Choosing a grid of initial conditions in the (θ, x) plane, we can compute the basins of attraction of each of these attractors and visualize the structure of these basins. This is shown for one part of the state space close to the interior crisis in Fig. 6.10. The white dots denote points belonging to the basin of the plotted attractor (bold line), while the grey dots mark the union of the basins of the two other, not plotted attractors. We obtain a fractal basin boundary, whose structure is rather complex and the open neighborhood

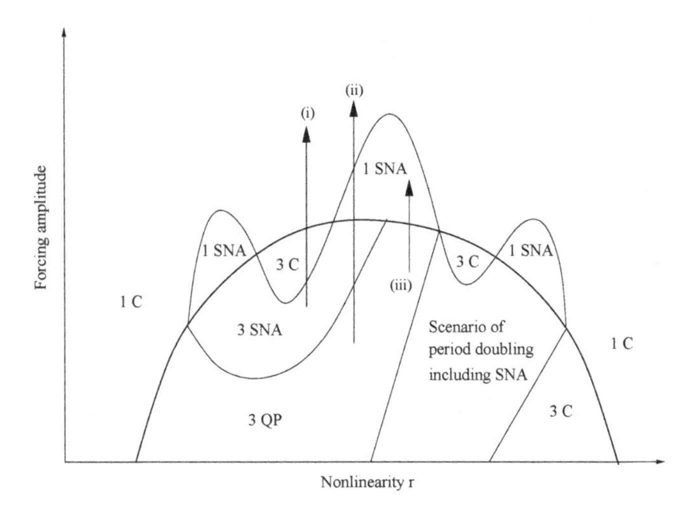

Fig. 6.8 Sketch of the dynamical regimes in the period-3 window of the quasiperiodically forced logistic map. The dynamical regimes are denoted by QP (quasiperiodic), SNA, and C (chaotic). The numbers in front correspond to the number of branches (pieces) of the respective attractor.

about the attractor contains islands of other basins of attraction. As we approach the interior crisis point with increasing forcing amplitude ε, the torus gets closer to the basin boundary (Fig. 6.10). Finally it collides with its basin boundary, or more precisely with a chaotic saddle embedded in it, to form a one branch SNA (Fig. 6.9b). The interior crisis yields a large strange nonchaotic attractor due to a collision of a smooth torus with a fractal basin boundary. Thus the third kind of interior crisis in quasiperiodically forced systems is not a chaos-chaos transition. Besides the sudden enlargement in the size of the attractor a *transition between smoothness and strangeness* in the structure of the attractor occurs. The former unstable chaotic saddle becomes embedded in the SNA and gives rise to local regions of expansion of nearby trajectories.

The temporal behavior beyond the interior crisis can be characterized as crisis-induced intermittency. A trajectory of the third iterate of $x_{n+1} = rx_n(1 - x_n)$ spends some long stretch of time in the vicinity of one of the former attractors, then it bursts out from this region and bounces around in the region of the former chaotic saddle until it comes close to the same or another former attractor where it remains again for some time interval. This way the trajectory is jumping irregularly between the three former attractors as time goes to infinity. This temporal behavior, which can be

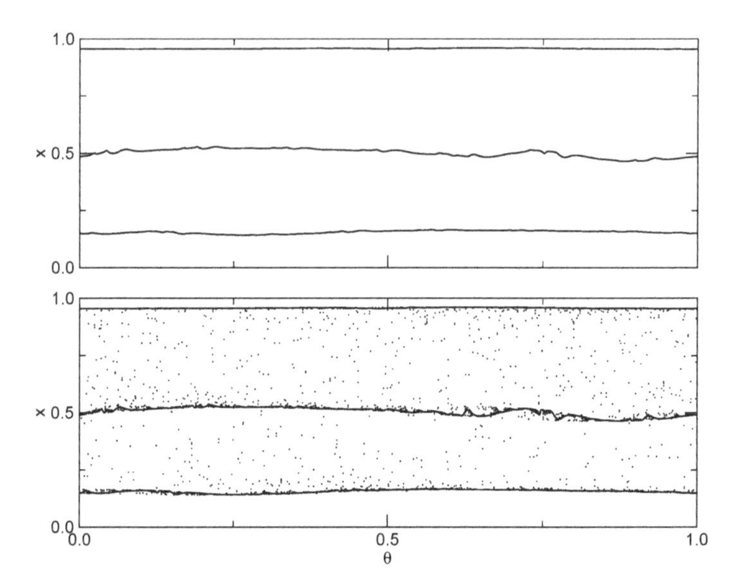

Fig. 6.9 Emergence of an SNA due to an interior crisis in the quasiperiodically forced logistic map ($r = 3.836$): (top panel) prior to the interior crisis for $\varepsilon = 0.0023$; (bottom panel) SNA beyond the interior crisis for $\varepsilon = 0.00242$.

observed for different types of crises, can be described by a specific mean time $\langle\tau\rangle$ corresponding to the type of the crisis [Grebogi et al. 1987a]. For the interior crisis this mean time $\langle\tau\rangle$ is defined as the average over the time intervals between bursts for a long orbit. The time intervals between two subsequent bursts seem to be more or less random. But their average value $\langle\tau\rangle$ possesses a scaling law as the forcing strength ε approaches the critical value ε_c where the crisis occurs.

$$\langle\tau\rangle \sim (\varepsilon - \varepsilon_c)^{-\gamma} \tag{6.4}$$

We have checked whether such a power law behavior can be obtained also for the interior crisis in quasiperiodically forced systems. Indeed we find a power law (Fig. 6.11 for $r = 3.836$). The critical exponent γ for this parameter value has been determined as $\gamma \approx 0.45$. This exponent is similar to the exponent obtained for the unforced logistic map. In general, the determination of the scaling behavior turned out to be rather difficult since the scaling regions for fitting the exponent γ are rather small, in most cases much smaller as the one shown in Fig. 6.11. Therefore, the influence of the quasiperiodic forcing on the scaling exponent is not so obvious and needs further investigation.

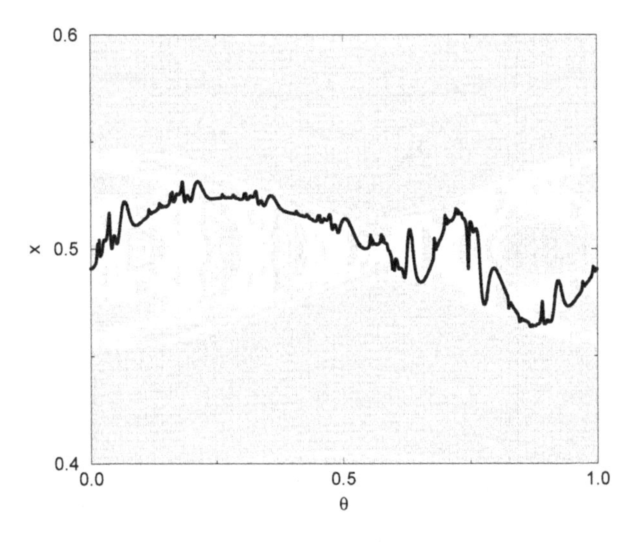

Fig. 6.10 Visualization of the interior crisis leading to an SNA in the quasiperiodically forced logistic map ($r = 3.836, \varepsilon = 0.002418$). The white dots denote the basin of attraction of the attractor plotted (bold line) while the dark grey points indicate the points belonging to the basins of two other torus attractors which are outside the part of the state space shown.

6.2.5 *Boundary crisis*

Bifurcations as sudden changes in the dynamical behavior of a system can not only lead to the appearance of certain attractors but also to their disappearance. The most important bifurcation which is related to such a disappearance of an attractor is the boundary crisis [Grebogi et al. 1982]. Its mechanism is based on a collision of the attractor with the boundary of its own basin of attraction. As a result the attractor is transformed to a chaotic saddle which can be observed in simulations as a chaotic transient right beyond the boundary crisis, while the long-term behavior is dominated by another attractor. Boundary crisis is typical for systems with multiple coexisting attractors [Feudel et al. 1996a]. However, they also occur in many other model systems, which possess besides the physically relevant attractor a second attractor at infinity. In the latter case those models have a restricted validity, which ends as soon as the boundary crisis occurs and all trajectories escape to infinity and the model becomes divergent.

The boundary crisis has been studied mainly in chaotic systems, where the chaotic attractor collides with its basin of attraction. For the famous logistic map the boundary crisis occurs at $a = 2$ for (6.1) or $r = 4$ for

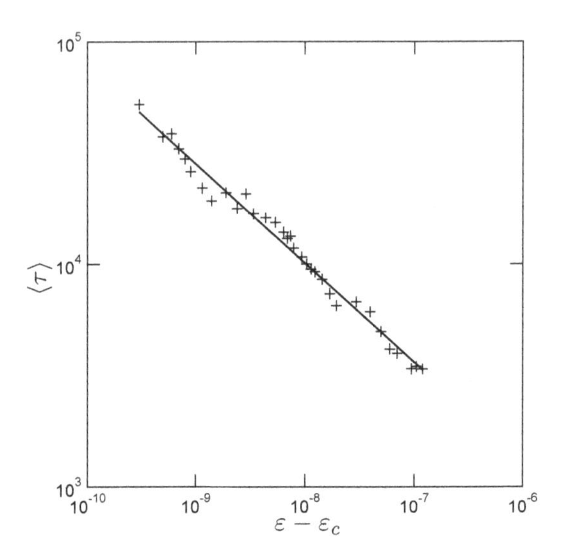

Fig. 6.11 Scaling behavior for crisis-induced intermittency for $r = 3.836$.

(6.3), respectively. At this parameter value the attractor collides with the unstable periodic orbit which is the continuation of the orbit appearing in the saddle-node bifurcation at $a = -0.25$ or $r = 1$. Beyond the boundary crisis every trajectory escapes to infinity and the map is no longer confined to the unit interval. A detailed theory in terms of stable and unstable manifolds which are involved in the crisis has been developed for the Hénon map and can be found in many text books on chaos theory [Ott 1992; Alligood et al. 1997; Argyris et al. 1994].

According to the aim of this book we consider here only the influence of a quasiperiodic forcing on the boundary crisis. This has been studied by Osinga and Feudel [2000] using the Hénon map as a paradigmatic example. Since these computations are rather involved we restrict ourselves here to a simpler example to give only a flavor of the bifurcations happening in the neighborhood of a boundary crisis. For this purpose we employ again the forced quadratic map where several routes for the disappearance of the attractor are possible which qualitatively correspond to the transitions observed in the Hénon map. Looking at the dynamical behavior in parameter space (Fig. 6.1) we notice the large region denoted by Dv meaning divergence. The boundary of this region is given by the boundary crisis of the attractor, which can be either chaotic (along the dark grey region) or regular (quasiperiodic or strange nonchaotic along the light grey region). We note further that the line of boundary crisis of the chaotic attractor is a rather straight line while the line of boundary crisis of the regular attractor forms some kind of bubble. We discuss two different boundary crises, whose routes are denoted by D and E in the (a, ε) parameter plane (Fig. 6.1).

The route along the arrow D corresponds to the usual boundary crisis. Prior to the crisis we find a chaotic attractor which collides with the boundary of its basin of attraction and disappears. Fig. 6.12 (top panel) shows the attractor and its basin just before the collision. We note that the boundary of the basin of attraction is a smooth curve, so that in the crisis, a *fractal attractor collides with a smooth boundary.*

More interesting is the route along arrow E. Here the attractor is a smooth torus before the crisis, but we find that the basin of attraction is fractal (Fig. 6.12 bottom panel). Thus in the second kind of boundary crisis a *smooth attractor collides with a fractal basin boundary.* This type of crisis can be observed only along a part of the boundary crisis line around the bubble. Whether the attractor is still smooth in the boundary crisis or it looses its smoothness prior to or directly in the transition is still an open question.

To find out the connection between these apparently two different kinds of boundary crises let us draw attention to the points where two different boundary crisis lines meet. There are two such points, the first on one side of the bubble and the second one the other side. These two points correspond to *double crisis points* which have been studied in several nonlinear dynamical systems [Gallas et al. 1993; Stewart et al. 1995; Osinga and Feudel 2000; Feudel and Grebogi 2003]. Such a double crisis is characterized by a simultaneous occurrence of a boundary crisis, where the attractor touches its basin boundary and an interior crisis, where the attractor changes suddenly its size. For the same parameter value a metamorphosis of the basin boundary takes place, where a sudden change in the structure of the basin boundary is observed. A sketch of such a double crisis vertex point is shown in Fig. 6.13, showing the four bifurcation lines which meet in the vertex.

The two boundary crisis lines are distinguished by the different kinds of the attractors (chaotic and regular, respectively) which are involved in the boundary crisis. The interior crisis line is in our case the line which separates the quasiperiodic motion from the strange nonchaotic and chaotic motion. The transitions crossing this bifurcation line have been studied in detail by Venkatesan et al. [2000]. They called them intermittency transition (cf. Section 6.5 and arrow F in Fig. 6.1). In the light of the study of double crisis vertices this intermittency transition can be interpreted in

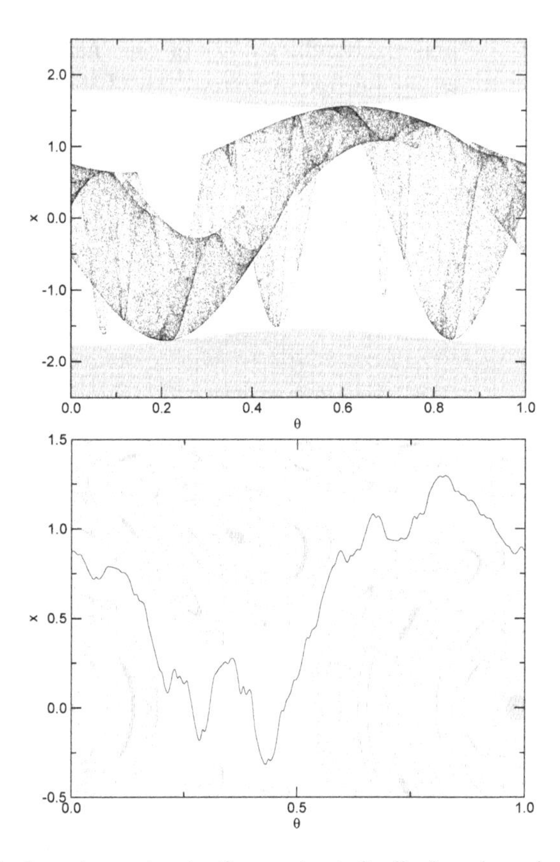

Fig. 6.12 Basin boundary crises in the quasiperiodically forced quadratic map. The attractor is shown in black. There are two basins of attraction. White points denote the basin of attraction of the attractor plotted, grey points go to infinity. Top panel: chaotic attractor and smooth basin boundary at $a = 1.1, \varepsilon = 0.46$; bottom panel: torus attractor and fractal basin boundary at $a = 1.2, \varepsilon = 0.42$.

terms of an interior crisis. It has been shown in [Osinga and Feudel 2000] for the forced Hénon map that the two boundary crisis lines corresponding to the boundary crisis with the chaotic attractor are connected by a line of basin boundary bifurcation. Following the same ideas we can argue that this basin bifurcation happens also in the logistic map. This basin boundary bifurcation is indicated by a dashed line in parameter space (Fig. 6.1).

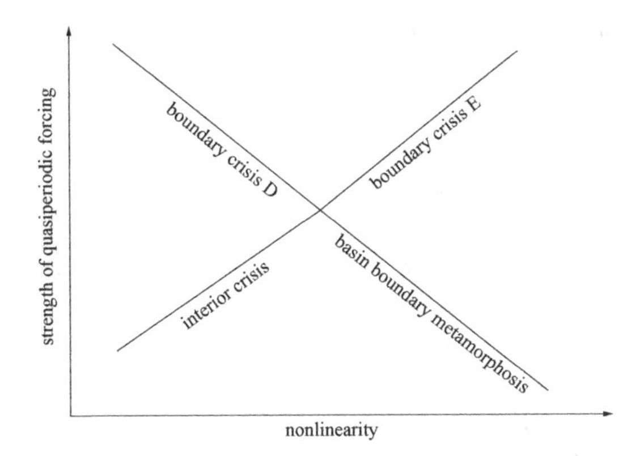

Fig. 6.13 Sketch of a double crisis vertex, where four bifurcation lines meet. The two boundary crisis lines denote on one hand the boundary crisis along route D in Fig. 6.1 where the fractal attractor collides with a smooth boundary while on the other hand the boundary crisis along route E , where the smooth attractor collides with a fractal basin boundary. Transitions crossing the interior crisis line correspond to the intermittency transition explained in Section 6.5.

6.2.6 *Basin boundary bifurcation*

Many physical, chemical, biological, and engineering systems are known to possess multiple coexisting final states. Multiple coexisting stable states mean that for a given set of parameter values, different choices of the initial condition $\mathbf{x}_0$ can lead to distinctly different asymptotic attractors, each with its own basin of attraction. The basin of attraction of an attractor is the set of initial conditions in the phase space that asymptote to this attractor. The boundaries that separate different basins of attraction are the *basin boundaries*, which can be either smooth or fractal [Grebogi et al. 1983a; McDonald et al. 1985; Moon 1984; Grebogi et al. 1988; Iansiti et al. 1985; Gwinn and Westervelt 1986; Grebogi et al. 1986a; Park et al. 1989]. To distinguish between smooth and fractal boundaries one can use the concept of the box counting dimension which gives an estimate of the fractal dimension of a geometrical object [Ott 1992]. When the boundary is smooth, its box or capacity dimension D_0 (cf. Chapter 4) is one less than that of the state space, *i.e.*, $D_0 = N - 1$. For a fractal basin boundary, its dimension D_0 is a fractional number that satisfies $(N-1) < D_0 < N$. More recently, another kind of basin boundaries has been identified. These so-called *Wada basin boundaries* are fractal basin boundaries which separate more than two basins of attraction. In the most complex case all attractors coexisting in a nonlinear dynamical system possess the same basin boundary. Such Wada basins have been studied in the dynamics of different nonlinear systems [Kennedy and Yorke 1991; Nusse and Yorke 1996; Poon et al. 1996].

Fractal basin boundaries are a robust property of nonlinear dynamical systems. It can be expected that all systems which possess fractal basin boundaries without forcing like the Duffing oscillator [Ueda et al. 1990] or the Hénon map [Grebogi et al. 1986b, 1987b], exhibit this property also under quasiperiodic forcing. Thus it is not a surprise that fractal and Wada basin boundaries have been found in quasiperiodically forced systems as well [Feudel et al. 1998a; Osinga and Feudel 2000] (cf. Section 6.2.5). In the following we will show how fractal basin boundaries look like in quasiperiodically forced systems. Moreover, we will discuss a basin boundary bifurcation, also called basin boundary metamorphosis, which is due to the influence of quasiperiodic forcing and therefore, specific to this system class. To simplify our considerations we explain the influence of quasiperiodic forcing on fractal basin boundaries using maps. The same kind of phenomena can be found in flows as well.

Let us consider the following class of dynamical systems

$$\mathbf{x}_{n+1} = \mathbf{F}(\mathbf{x}_n) + \mathbf{M}(\theta_n) \,, \tag{6.5}$$

$$\theta_{n+1} = \theta_n + w \pmod 1 \,, \tag{6.6}$$

where $\mathbf{F}(\mathbf{x})$ is a nonlinear map that can exhibit multiple coexisting attractors, and $\mathbf{M}(\theta)$ models the external quasiperiodic driving. Under certain conditions such maps can exhibit fractal boundaries between the basins of attraction of the different coexisting attractors. Suppose that we consider the map $\mathbf{F}$ in such a parameter region where the basin boundaries are fractal in the unforced case ($\mathbf{M} = 0$) and apply now a quasiperiodic forcing. For small forcing amplitudes we find that the fractal basin boundaries persist and their structure is the fractal set of the unforced map $\mathbf{C}$ cross the unit circle in θ direction ($\mathbf{C} \times \mathbf{S}^1$). Its box-counting dimension is equal to the dimension of the fractal set $\mathbf{C}$ in the unforced case D_0^u enlarged by one for the dimension of the unit circle: $D_0 = D_0^u + 1$. Due to increasing quasiperiodic forcing another structure of the basin boundary becomes possible which destroys the $C \times S^1$ structure and leads to a basin boundary which is characterized by a box dimension $D_0 > D_0^u + 1$. Finally, for large forcing amplitudes the basin boundary fills most of the state space and its box dimension is close to the dimension of the state space. In the following we demonstrate this basin boundary bifurcation destroying the $\mathbf{C} \times \mathbf{S}^1$ structure and explain how this bifurcation can occur.

As already mentioned we start our study with a map where in the unforced case already fractal basin boundaries between the attractors exist. We choose $\mathbf{F}(\mathbf{x})$ in (6.5) to be the three-times iterated version of the logistic map, that is, $F(x) = f^{(3)}(x)$ with $f(x) = rx(1-x)$. For simplicity we choose the parameter r in the logistic map $f(x)$ in such a way that the dynamics is in the period-3 window and, hence, the map $F(x)$ possesses three isolated attractors,namely fixed points, each with its own basin of attraction. We choose the parameter r so that these attractors are fixed-point attractors in the unforced case. Furthermore, we consider the simplest type of driving: $M(\theta_n) = \varepsilon \cos(2\pi\theta_n)$, where ε is the driving amplitude. The driving frequency w in (6.6) is chosen to be the inverse of the golden mean: $\omega = (\sqrt{5}-1)/2$. When the quasiperiodic forcing is present ($\varepsilon \neq 0$), the resulting map possesses three isolated attractors in the two-dimensional phase space (θ, x) which can be either tori, strange non-chaotic or chaotic attractors depending on the strength of the forcing (cf. Section 6.2.4. Without forcing the boundaries between the basins of attraction of the three fixed points are Cantor sets (fractal) [McDonald et al. 1985; Grebogi et al. 1988; Napiórkowski 1986]. Under the influence of small amplitude forcing these boundaries are topologically fractal boundary sets that already exist in the map $F(x)$ cross the circle (in the θ direction) as shown in Fig. 6.14a. When varying the system parameters, the structure of the basin boundaries changes suddenly. Instead of the basin boundary consisting of a Cantor set of invariant circles ($\mathbf{C} \times \mathbf{S}^1$), another type of basin boundary arises. In this type, the basins of attraction of one attractor have isolated "islands" immersed in the basins of the other attractors. Fig. 6.14c shows an example for such a basin. Panel (b) shows the basin of attraction just beyond the basin boundary metamorphosis, where the islands are still rather small and scattered in the basin of attraction of the other attractor. Panel (c) shows the picture already far beyond the basin boundary bifurcation. The formation of those islands is the result of a basin boundary bifurcation induced by the quasiperiodic driving and emerging at particular values of the system parameters.

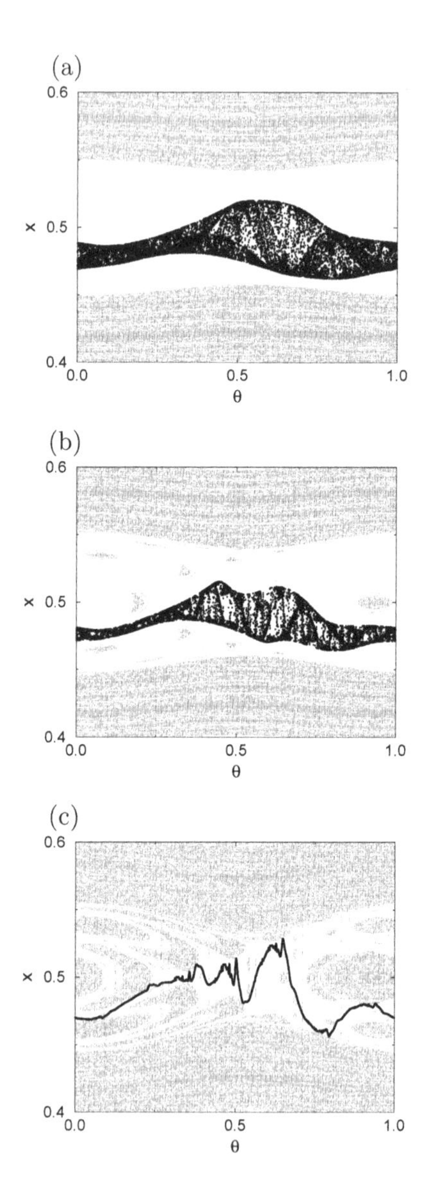

Fig. 6.14 Basin boundary metamorphosis in the quasiperiodically forced logistic map: The white dots denote the basin of attraction of the attractor plotted (black points) while the grey dots indicate the points belonging to the basins of attraction of the two other not shown attractors ($r = 3.846$): (a) prior to the bifurcation ($\varepsilon = 0.001$); (b) just beyond the bifurcation point ($\varepsilon = 0.0015$); (c) far away from the basin boundary bifurcation ($\varepsilon = 0.0024$).

We now characterize, quantitatively, the fractal basin boundary in Figs. 6.14. It has been known that fractal basin boundaries pose a fundamental difficulty in the prediction of the asymptotic attractor of the system [Grebogi et al. 1983c,a] because of the interwoven fractal structure of the basins of attraction and because of the inevitable error in the specification of initial conditions and system parameters. This is called the *final state sensitivity* [Grebogi et al. 1983a]. Let ϵ be such an error. Then the probability for two initial conditions, being ϵ distance apart, to asymptote to different attractors scales with ϵ as,

$$P(\epsilon) \sim \epsilon^{\alpha}, \tag{6.7}$$

where the scaling exponent is the uncertainty exponent α, with $0 < \alpha \leq 1$ [Grebogi et al. 1983a]. We obtain $\alpha \approx 0.11$ for the union of the basin boundaries in Fig. 6.14b and $\alpha \approx 0.05$ for the basin boundaries in Fig. 6.14c. The relation between the uncertainty exponent α and the box counting dimension D_0 ($\alpha = N - D_0$ where N is the dimension of the state space) allows the estimation of D_0 from measurements of α [Grebogi et al. 1988]. In our example, the box counting dimension of the union of the basin boundaries between the three basins is then $D_0 = 2 - \alpha \approx 1.89$ (panel (b)) and $D_0 \approx 1.95$ (panel (c)), respectively, which is close to the phase-space dimension. This indicates that the basin boundary separating the three attractors has an arbitrarily fine-scale structure. Moreover, one can show for this example that all three basins of attraction have a *common* boundary and thus possess the Wada property [Feudel et al. 1998b].

Next we address the following questions: How such basin boundary bifurcations can occur in quasiperiodically driven systems and what are the unique characteristics of such bifurcations? To gain some insight, we refer to Fig. 6.15, the plot of the one-dimensional map $F(x)$ without driving. In the figure, there are three square regions in which the three fixed point attractors lie. The one-dimensional subintervals $A1$, $A2$, and $A3$ belong entirely to the basins of the three attractors. The boundary between the three basins of attraction must then lie in the one-dimensional set which is the complement set of the subintervals $A1$, $A2$, and $A3$ in $[0, 1]$. Concentrating on one of the complement intervals, say $[a, b]$, we see that there are three subintervals in $[a, b]$, denoted by 1, 2, and 3, which are the preimages of $A1$, $A2$, and $A3$, respectively. The interval $[a, b]$ thus contains all three basins and contains the complement set of the joint set of subintervals 1, 2, and 3 in $[a, b]$. This complement set consists of four subintervals, denoted by $S1$, $S2$, $S3$, and $S4$, respectively, as shown in Fig. 6.15. Now look at one

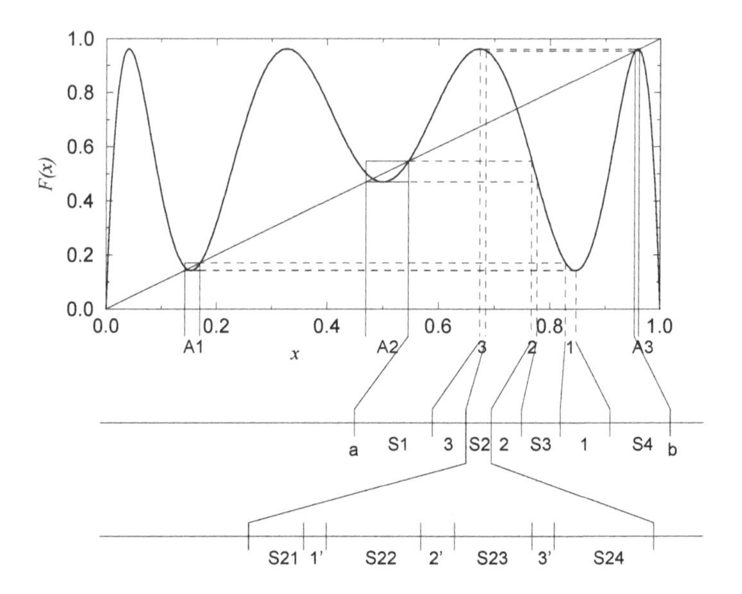

Fig. 6.15 The third iterate of the logistic map with no forcing.

of these four subintervals, say $S2$, the one in between 3 and 2. We see that there are three subintervals in $S2$, denoted by $1'$, $2'$, and $3'$, respectively, which map to $A1$, $A2$, and $A3$ in two iterations. The subinterval $S2$, which is smaller than the original interval $[a, b]$, contains all three basins. In a similar fashion, it is easy to see that there are four still smaller subintervals $S21$, $S22$, $S23$, and $S24$ in $S2$ that contain the basin boundary. Any of these smaller subintervals must contain all three basins. By examining the nth preimages of the subintervals $A1$, $A2$, and $A3$ in the limit $n \to \infty$, we see that an arbitrarily small subinterval Snj $(j = 1, 2, 3, 4)$ must contain all three basins. The boundary between the three basins must then be unique, fractal, and Wada.

Now imagine we turn on the quasiperiodic forcing. At small forcing amplitude, the critical points are still in the square so that the subintervals A_1, A_2, and A_3 are still open basins of the three attractors. At different locations of θ, the driving is different. Thus, the lengths of the subintervals A_1, A_2, and A_3 are different for different θ values but, nonetheless, the lengths change smoothly due to the smooth driving function $\varepsilon \cos 2\pi\theta$ used, as shown in Fig. 6.14a by the large white region about $x = 0.5$. As the forcing amplitude increases, at some locations of θ the driving term $\varepsilon \cos 2\pi\theta$ is larger so that at these locations, the critical points of the map $F(x)$ are no longer contained in the squares. When this happens, a subinterval, say A_2, contains part of the basins of the attractors that are in A_1 and A_3. In this sense, the basin of the attractor in A_2, which is originally connected, now invades the basins of the other attractors. In the two-dimensional phase space (θ, x), we then expect to see complicated basin structures in the originally open basins. In particular, since the effect of forcing is different at different θ values, the newly created basins in the originally open basins form an "island" structure, as shown in Fig. 6.14c.

This island-like basin structure created after a basin boundary bifurcation or metamorphosis is a unique feature of systems, where $\mathbf{F}$ in (6.5) is noninvertible. If $\mathbf{F}$ is invertible, the basin bifurcation will also occur but look a bit different. The common feature of the basin boundary bifurcation in noninvertible and invertible systems will be the discontinuous change in the box dimension of the basin boundary which reflects the change from a structure $\mathbf{C} \times \mathbf{S}^1$ to a structure which does not exhibit this property.

Finally we show for our example quantitatively that the bifurcation destroys the $\mathbf{C} \times \mathbf{S}^1$ structure of the basin of attraction. To this end we compute the box dimension of the union of all basin boundaries depending on the bifurcation parameter. As we discussed above, the basin boundary in Fig. 6.14a is a Cantor set of invariant circles ($\mathbf{C} \times \mathbf{S}^1$) and thus, their box dimension is the sum of the dimension of the Cantor set obtained from the unforced logistic map and one, the dimension of the invariant circles, $D_0 = D_0^u + 1$. To monitor the box dimension when approaching the basin boundary bifurcation, let us fix $r = 3.846$ and vary the forcing amplitude ε. While increasing ε we find that the uncertainty exponent α and, consequently, the box dimension of the boundary D_0 is independent of ε (Fig. 6.16) prior to the bifurcation. Within the accuracy of our computations, which is measured by the standard deviations of the least square fits for each α, we obtain $\alpha = \alpha(\varepsilon) = const$ and subsequently $D_0 = D_0(\varepsilon) = const$ [curve (a)] for the basin boundary which is a Cantor set of invariant circles as in Fig. 6.14a. Beyond the basin boundary bifurcation which occurs at $\varepsilon_c \sim 0.001424$ the uncertainty exponent α appears to depend linearly on the forcing amplitude ε. The decrease in α corresponds to an increase in the box dimension D_0 of the basin boundary [curve (b) in Fig. 6.16], meaning that the box dimension becomes larger than $D_0^u + 1$. Finally, we can expect that the box dimension of the basin boundary approaches the dimension of the state space when increasing the amplitude of forcing further. This linear dependency can be understood using the same arguments as in [Park et al. 1992] where the authors analyze a basin bound-

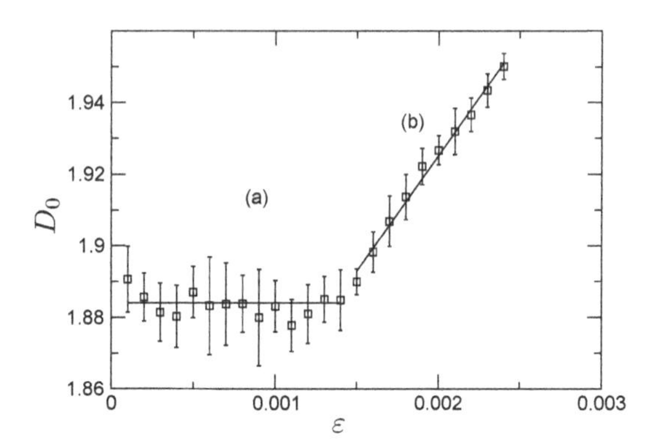

Fig. 6.16 Box dimension D_0 depending on the amplitude of forcing ε for $r = 3.846$.

ary bifurcation in a piecewise linear, noninvertible map. This map is similar to our map shown in Fig. 6.15 but instead of the parabola-like functions in the small rectangles they have considered piecewise linear functions. The bifurcation occurs as soon as the tips cross the boundary of the rectangle [Devaney 1989]. An analytical study shows that in general one obtains a power law dependence of the dimension of the basin boundary on the bifurcation parameter ε: $(D_0 - (D_0^u + 1)) \sim (\varepsilon - \varepsilon_c)^\gamma$, where $D_0^u + 1$ stands for the dimension of the boundary prior to the basin boundary bifurcation and ε_c denotes the forcing amplitude at the basin boundary bifurcation point. However, for very small forcing amplitudes ε applied in our example, this dependence is essentially linear as observed in the numerical experiments.

6.3 Bifurcations in the quasiperiodically forced circle map

Another important route to chaos besides the period doubling route scenario is the route via quasiperiodicity, or Ruelle-Takens route [Ruelle and Takens 1971]. This route is related to the destruction of tori T^3 since they are structurally unstable. If certain perturbations are applied to the model system, then the quasiperiodic motion will turn into chaotic motion and as a result the torus will be destroyed [Afraimovich and Shilnikov 1983; Anishchenko et al. 1985, 1994]. This scenario can be conveniently studied by circle maps which describe the dynamics of the phase angle in the Poincaré section (cf. Chapter 2):

$$\varphi_{n+1} = \varphi_n + f(2\pi\varphi_n) \pmod 1 . \tag{6.8}$$

In this section we restrict our attention to one-to-one maps, i.e. those with a monotonic dependency of φ_{n+1} on φ_n. In the case when map (6.8) has maxima and minima, it is locally similar to the logistic map, so that the considerations of the previous section apply.

As already explained in Chapter 2, the dynamics of circle maps may be characterized by the rotation number ρ_φ. Intuitively, the rotation number measures the average angular velocity of an orbit around the circle $\{\varphi \pmod 1\}$. The unforced circle maps possess in general two types of dynamics: two-frequency quasiperiodic motion on a torus and phase-locked, periodic motion. Both types of dynamics can be distinguished by computing the rotation number ρ_φ defined by (2.21). If the rotation number is rational, i.e. $\rho_\varphi = p/q$ with p and q integers, then the corresponding motion is phase-locked. In case the rotation number is irrational, then the motion is quasiperiodic. The parameter space of such models, usually spanned by a nonlinearity and a phase shift parameter, is divided into regions with rational and irrational rotation numbers.

The most prominent example for a circle map is the famous Arnold sine map which is defined as:

$$\varphi_{n+1} = \varphi_n + c + a\sin(2\pi\varphi_n) \pmod 1\ , \tag{6.9}$$

The parameter c is the phase shift, while a corresponds to the strength of the nonlinearity of the map.

In case of the Arnold sine map the regions with phase-locked motion are called Arnold tongues. In each Arnold tongue corresponding to a certain rotation number $\rho_\varphi = p/q$ there exist at least two periodic orbits of period q, one stable and one unstable. The boundaries of the Arnold tongues correspond to saddle-node bifurcations of these q-periodic orbits. In the parameter space spanned by phase shift and nonlinearity (c, a) the tongues increase monotonously in width as the nonlinearity increases. The left boundary of each tongue is related to the appearance of the stable and the unstable q-periodic orbits, while the right boundary marks their disappearance due to annihilation of the two invariant sets. These saddle-node bifurcations are also called phase-lockings since they denote a transition from quasiperiodic motion on a torus to phase-locked periodic motion on the same torus.

6.3.1 *Smooth saddle-node bifurcation of tori*

Let us now discuss what happens to the phase-locking bifurcation when a quasiperiodic forcing is applied. To simplify the considerations we will discuss the bifurcations using the forced sine map:

$$\varphi_{n+1} = \varphi_n + c + a\sin(2\pi\varphi_n) + \varepsilon\sin(2\pi\theta_n) \pmod 1 , \tag{6.10}$$

$$\theta_{n+1} = \theta_n + \omega \pmod 1 . \tag{6.11}$$

We focus here on the parameter range $a < (2\pi)^{-1}$ where the sine map is invertible, i.e. chaos is ruled out. Invertibility allows us to compute the stable as well as the unstable tori.

As already outlined in the beginning of this chapter we expect to find a phase-locking bifurcation corresponding to a saddle-node bifurcation of tori, which is the quasiperiodic equivalent to the saddle-node bifurcation of periodic orbits in the unforced case. Indeed, we find such a bifurcation where the two tori (stable and unstable) existing inside the tongue approach each other when the parameters are changed in such a way that the system moves towards the boundary of the tongue. Exactly on the boundary of the tongue the two tori annihilate each other. We call this a *smooth* saddle-node bifurcation where both, the stable and the unstable torus, meet in *all* θ values. This corresponds to a *merging* of the two tori resulting in a three-frequency quasiperiodic motion which fills the whole torus (Fig. 6.17). Thus the transition from phase-locked periodic motion to quasiperiodic motion in unforced systems turns into a transition from two-frequency quasiperiodic motion on T^2 to three-frequency motion on T^3. However, it turns out that this kind of bifurcation can be obtained only for small forcing amplitudes. As the forcing amplitude increases, the tori inside the Arnold tongues become more and more wrinkled and the bifurcation becomes more complex, giving rise to the formation of strange nonchaotic attractors as discussed in the next subsection.

6.3.2 *Non-smooth collision of a stable and an unstable torus*

One of the basic mechanisms of the emergence of SNA is a *non-smooth* collision of a stable and an unstable torus [Feudel et al. 1995a, 1997]. Since such collision phenomena take place in most of the transitions to SNA we will explain it here in detail. To study the appearance of strange nonchaotic attractors in circle maps it is sufficient to focus on the dynamics within

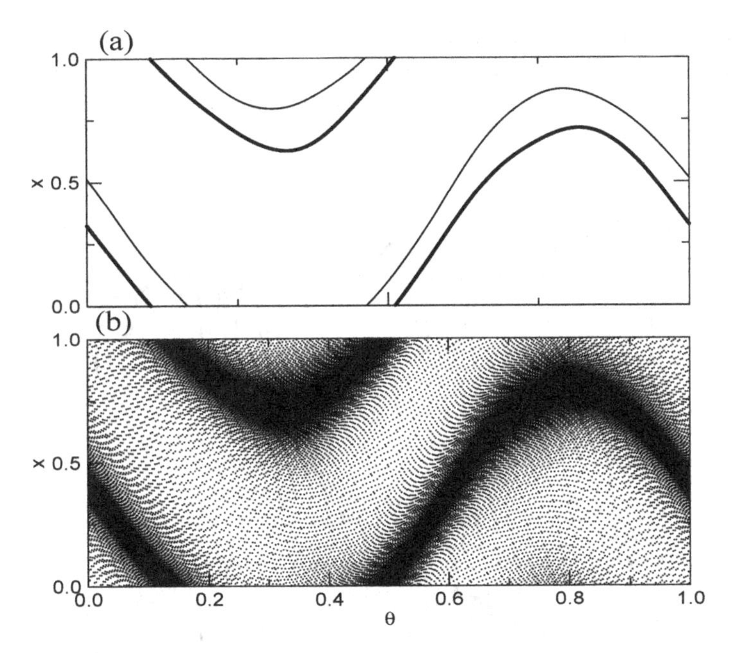

Fig. 6.17 Emergence of a three-frequency quasiperiodic motion due to the merging of a stable (bold line) and an unstable (thin line) torus in the quasiperiodically forced sine circle map ($a = 0.2, \varepsilon = 1.0$). (a) prior to the transition ($c = 0.01$); (b) beyond the transition ($c = 0.012$)

one Arnold tongue, say, in the tongue with rotation number zero. The mechanism will be the same in all other phase-locked regions with different rotation numbers.

In the quasiperiodically forced sine circle map we find a stable and an unstable two-frequency quasiperiodic motion for zero rotation number. As long as the forcing amplitude is small, the stable and the unstable torus are smooth and far away from each other. With increasing forcing the two tori become more and more wrinkled. As a consequence they are far apart for some θ values and very close for other θ values (Fig. 6.18a). At a critical forcing amplitude ε_c the stable (bold line) and the unstable (thin line) torus *collide in a dense set of θ values.* Indeed, if the stable and the unstable tori have a common point (θ_0, φ_0), then because the tori are invariant sets, all the iterations of this point are also points which belong to both tori. But the trajectory of θ_0 is dense on the interval $[0, 1)$, therefore the common points of both tori constitute a dense set. This observation holds for the smooth collision of two tori as well, there simply *all* points of two tori coincide at the bifurcation. For the non-smooth collision there are also points that do not belong to the two tori. One can conclude that the whole situation must be non-smooth: two smooth curves cannot touch in a dense set of points.

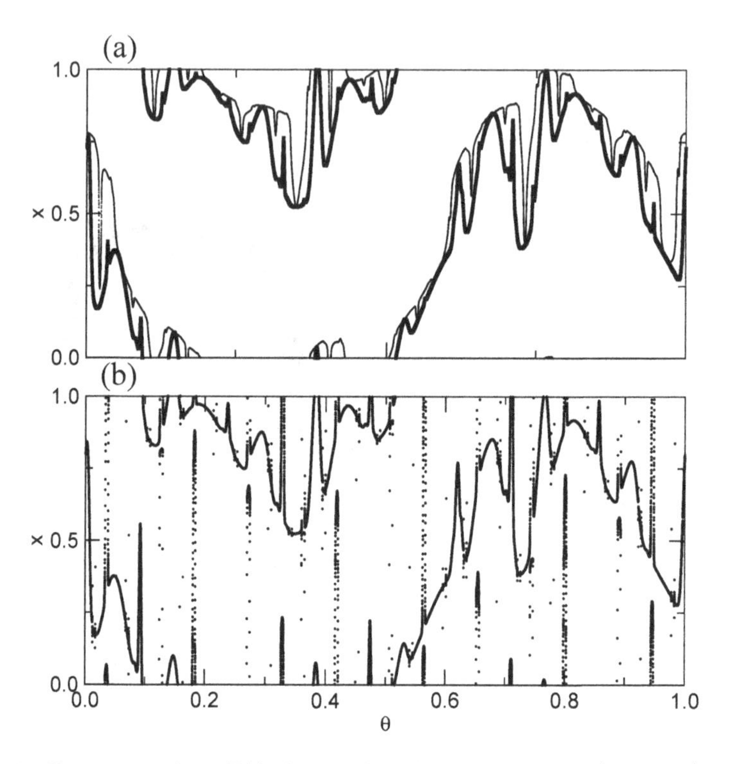

Fig. 6.18 Emergence of an SNA due to the collision of a stable (bold line) and an unstable torus (thin line) in the quasiperiodically forced sine circle map ($a = 0.9, \varepsilon = 1.0$). (a) prior to the transition ($c = 0.343$); (b) beyond the transition ($c = 0.345$)

The collision described leads to the emergence of a strange nonchaotic attractor (Fig. 6.18b). The strange geometry formed is due to the intertwined structure of the stable and unstable tori giving rise to regions with contraction and region with expansion of nearby trajectories. The unstable torus becomes embedded in the attractor and is thus responsible for the strangeness of the SNA. Since the collision happens only in a countable dense set of points in θ we call this transition a *non-smooth* saddle-node bifurcation of tori in contrast to the previously discussed smooth saddle-node bifurcation of tori shown in Fig. 6.17.

Note that in contrast to the bifurcation beyond a torus doubling (cf. Section 6.2.2) now both the stable and the unstable tori are wrinkled in the same way.

6.3.3 *Phase-locking regions under quasiperiodic forcing*

Both transitions, the smooth as well as the non-smooth saddle-node bifurcation of tori, described in the previous subsections, form the boundary of the Arnold tongue. While the smooth merging of the two tori is typical for small nonlinearities or small forcing amplitudes, the non-smooth collision of the tori in a dense set of points appears mainly for large nonlinearities or large forcings. Along the boundary of the tongue we find a critical point at which the smooth saddle-node bifurcation of tori turns into the non-smooth one. This point denotes the torus fractalization point (TF). A detailed analysis of the dynamics in the vicinity of this point and the scaling properties of the fractal attractor are discussed in Chapter 7.

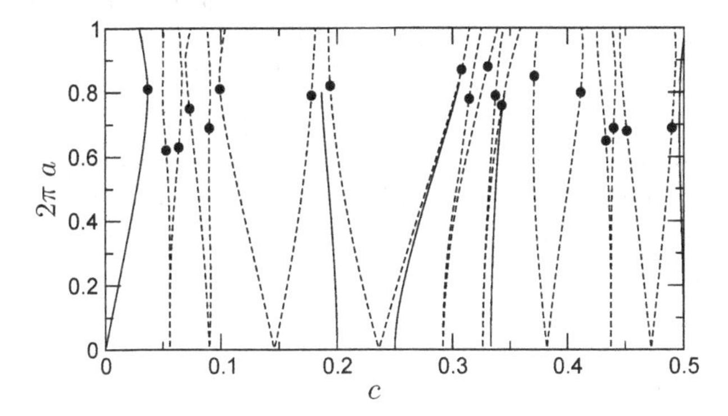

Fig. 6.19 Phase-locking regions in the forced sine circle map ($\varepsilon = 1.0$). The phase-lockings which are already present in the unforced case are denoted by solid lines while the phase-locking regions which appear newly due to the quasiperiodic forcing are denoted by dashed lines. From left to right the drawn phase-locking regions have the rotation numbers: 0, $5-8\omega$, $5-3\omega$, $2-3\omega$, $1/5$, $2-\omega$,$1/4$, $4-6\omega$, $7-4\omega$, $1/3$, $1-\omega$, $6-9\omega$, $4-2\omega$, $1/2$. The filled circles denote the torus fractalization points (TF point) on the boundaries of the phase-locking regions.

There is another important property of the phase-locking regions which is induced by the quasiperiodic forcing. While the Arnold tongues in the unforced sine circle map are monotonously increasing in width with increasing nonlinearity this is no longer the case in the quasiperiodically forced sine circle map. Due to the quasiperiodic forcing the number as well as the shape of the phase-locking regions change [Feudel et al. 1997; Glendinning et al. 2000; Glendinning and Wiersig 1999; Vasylenko et al. 2004]. This is illustrated in Fig. 6.19 for a number of tongues. In forced circle maps the resonance condition for phase-locking is different (see Chapter 2). Since there are two frequencies 1 and ω the rotation number for phase-locked motion equals

$$\rho_\varphi = \frac{p}{q} + \frac{r}{q}\omega \, . \tag{6.12}$$

As a consequence much more regions of phase-locked motion are observed in the (a, c) parameter space compared to the unforced case. To highlight this fact in Fig. 6.19, the phase-locked regions which are already present in the unforced case and which correspond to the resonance condition $\rho_\varphi = p/q$ are denoted by solid lines. The additional phase-locked regions which appear due to (6.12) with $r \neq 0$ are marked with dashed lines.

Moreover, due to the forcing the shape of the Arnold tongues changes. In contrast to the unforced case we find that the width of the tongues is not monotonously increasing with increasing nonlinearity but is again decreasing with even higher nonlinearities. The increase and decrease in width depends strongly on the forcing amplitude. The decrease in width can be also observed in the parameter region where the map is invertible. For certain forcing amplitudes this decrease in width is so strong that the phase-locking region pinches off, as shown in Fig. 6.20 below. For this reason strange nonchaotic attractors which are formed at the boundaries of the phase-locking region can have rational as well as irrational rotation numbers.

The pinching of the phase-locked regions has another important implication for the devil's staircase. The devil's staircase describes the rotation numbers depending on the phase shift c of the unforced Arnold circle map for fixed nonlinearity: $\rho_\varphi = H(c)$. Usually, the nonlinearity parameter is set to $a = (2\pi)^{-1}$. The length of the steps in the staircase denote then the width of the phase-locked regions corresponding to the different rotation numbers. In the forced circle map these steps become smaller and smaller for increasing forcing amplitudes due to the pinching effect. Finally, the devil's staircase will degenerate to a straight line [Ding et al. 1989a; Feudel et al. 1995a].

We notice that identifying the nature of tori collision is a relatively simple numerical task. We illustrate it for the tongue with zero rotation number shown in Fig. 6.20. First, the boundary is determined via calculation of

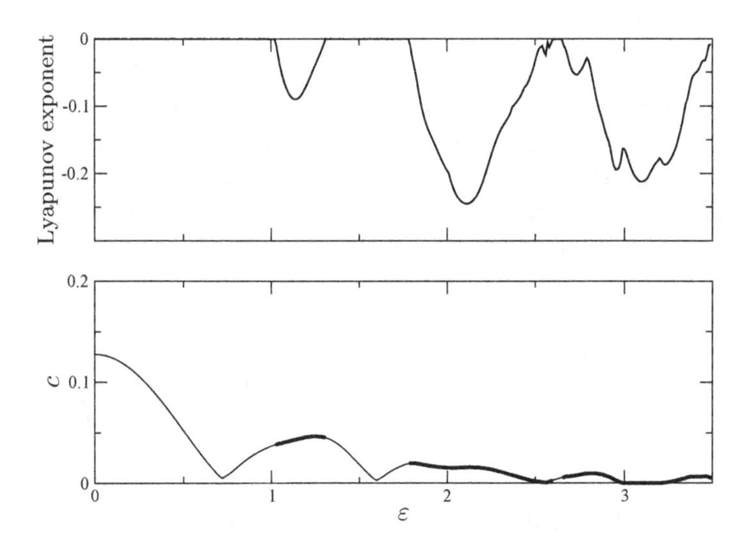

Fig. 6.20 Bottom panel: The boundary of the phase-locking region with zero rotation number in the (ε, c) plane depending on ε for fixed $a = 0.8$. Top panel: The Lyapunov exponent at this boundary. The pieces of the boundary where the Lyapunov exponent is negative correspond to the non-smooth transition to SNA (bold lines).

the rotation number according to (2.21) and checking if it is zero or not. In this way the border can be calculated with high precision. Next, one calculates the Lyapunov exponent at the border according to (2.10). For the smooth transition the stable torus coincides with the unstable one and both are marginally stable in linear approximation, thus the Lyapunov exponent vanishes. For the non-smooth transition the stable torus remains distinct from the unstable one (although they touch densely) and the Lyapunov exponent is negative. With this method we have constructed Fig. 6.20.

Let us now discuss this transition to SNA in terms of rational approximations, again assuming the case of the tongue with vanishing rotation number. If we approximate the external frequency ω by $\omega_k = F_{k-1}/F_k$, we get the following infinite set of maps:

$$x_{n+1} = x_n + c + a \sin 2\pi x_n + \varepsilon \sin[2\pi(n\omega_k + \theta_0)] \, . \tag{6.13}$$

For every k the system is periodically forced with period F_k and the resulting long-term dynamics depends on the initial phase θ_0. For each θ_0 we obtain an attractor and the union of these attractors for all $0 \leq \theta_0 < 1$ gives the $k-th$ rational approximation of the attractor in the quasiperiodically forced system. It is sufficient to consider only the $F_k - th$ iteration of the map to study the qualitative properties of the rational approximation. This way only one point of the periodic orbit with period F_k is studied. Furthermore, the investigation is restricted to the choice of θ_0 from the interval $[0, \theta_{max} = 1/F_k)$. Since we focus on nonlinearities $2\pi a < 1$ the map is invertible so that it is possible to compute stable as well as unstable fixed points and periodic orbits using forward and backward iteration.

The rational approximations for the quasiperiodic and strange non-chaotic attractors in the quasiperiodically forced circle map have been discussed in Chapter 3. For small ε the map $x_n \to x_{n+F_k}$ has a stable fixed point $x_{n+F_k} = x_n$ for all phases θ_0, if k is large enough. This gives the $k-th$ approximation of the stable torus. The backward iteration $x_n \to x_{n-F_k}$ has a stable fixed point as well corresponding to the approximation of the unstable torus.

Let us focus on transitions that occur for fixed ε as the parameter c, the phase shift, is varied. As c approaches the critical value, the distance between the sets of stable and unstable fixed points decreases, and at $c = c_{down}$ they touch at one point θ_{down}. For values of c that are slightly larger than c_{down} there still exist stable and unstable fixed points for some values of θ_0, but in a neighborhood of θ_{down} other regimes with a rational or irrational non-zero rotation number occur. A further increase of c leads to collisions of stable and unstable points for other θ_0, finally at c_{up} all pairs of stable and unstable fixed points disappear.

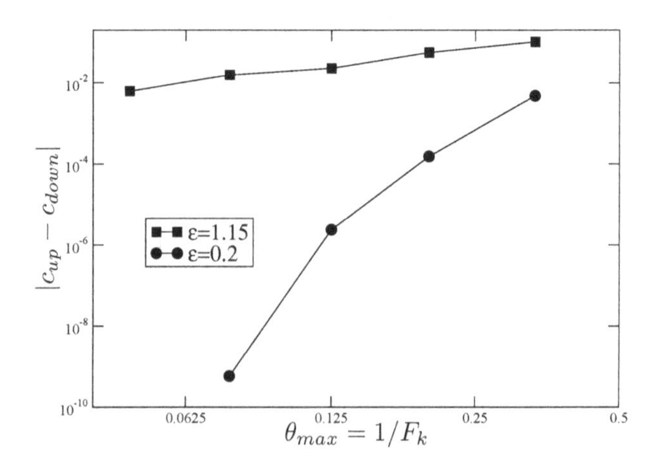

Fig. 6.21 Convergence of the bifurcation interval in rational approximations, for two values of parameter ε corresponding to smooth ($\varepsilon = 0.2$) and non-smooth ($\varepsilon = 1.15$) transitions, cf. Fig. 6.20.

The picture above holds qualitatively both for the smooth and non-smooth bifurcations. The main difference is in the quantitative properties. For the smooth tori merging the interval (c_{down}, c_{up}) tends to zero very fast – superexponentially – as the approximation number k increases, see [Chastell et al. 1995]. Contrary to this, for the non-smooth bifurcation the interval (c_{down}, c_{up}) decreases slowly, see Fig. 6.21.

6.3.4 *Non-smooth pitchfork bifurcation*

Another transition yielding a strange nonchaotic attractor is related to the class of unforced systems possessing a pitchfork bifurcation of fixed points [Sturman 1999a; Osinga et al. 2001]. The emergence of an SNA in the quasiperiodically forced counterpart of such systems follows the same rule as the saddle-node and period doubling bifurcation of tori (see previous discussion in this chapter). For small forcings or nonlinearities the pitchfork bifurcation appears as a *smooth* pitchfork bifurcation of tori. Looking at the bifurcation in reverse direction (from two stable and one unstable torus towards only one stable torus) all the three tori approach each other and merge *in each θ value* to form a single torus. With increasing forcing the tori become again wrinkled and the transition to SNA appears as a result of a *non-smooth collision* of the two stable tori with their unstable parent formed in the pitchfork bifurcation. The three tori touch each other only in a dense set of points at the bifurcation.

Let us consider as an example the quasiperiodically forced sine circle map (6.10, 6.11). We start our illustration with the unforced system just beyond the pitchfork bifurcation. Then we have the coexistence of two stable and one unstable fixed point exhibiting a certain symmetry. Applying a quasiperiodic forcing yields two stable tori and one unstable torus in the middle. Increasing the forcing amplitude leads again to a wrinkling of the tori. Fig. 6.22 (top panel) shows the two stable tori (bold lines) and the unstable one (thin line) just prior to the collision. Finally the stable and unstable tori collide in a dense set of θ values giving rise to the formation of an SNA with an unstable torus embedded in it (Fig. 6.22, bottom panel). Note that there is a second unstable torus visible in Fig. 6.22 which is not involved in the bifurcation analyzed here.

The dynamical behavior on the created SNA beyond the pitchfork bifurcation is characterized by an intermittent process as it is well-known from interior and band-merging crises of chaotic attractors [Grebogi et al. 1987a]. Since there have been two different stable tori prior to the collision with

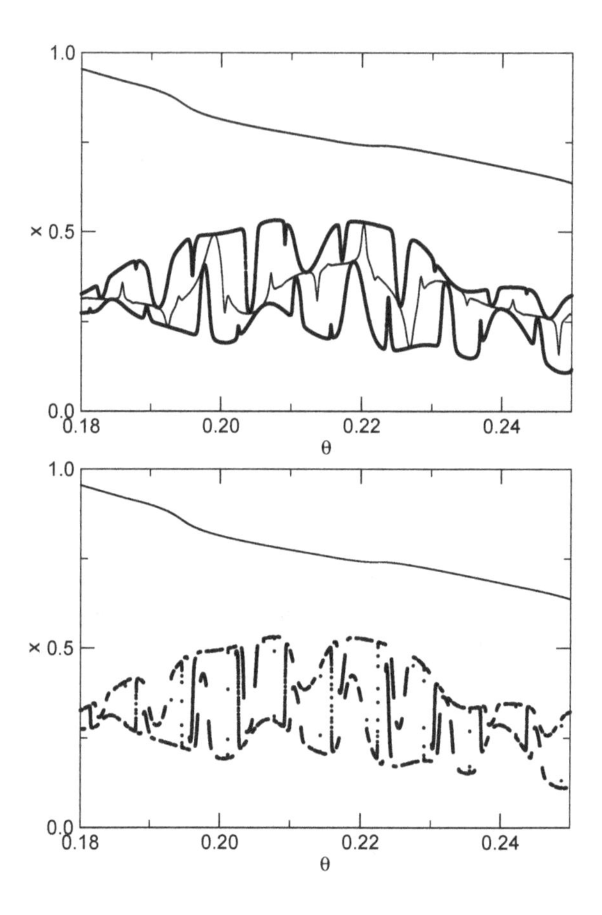

Fig. 6.22 Emergence of an SNA due to the collision of two stable tori beyond a pitchfork bifurcation (bold lines) with their unstable parent (thin line) in the quasiperiodically forced circle map ($a = 0.9, c = 0$). Top panel: just prior to the collision ($\varepsilon = 1.5675$); bottom panel: beyond the collision ($\varepsilon = 1.56765$).

the unstable torus, the remnant of these distinct solutions is still visible in the dynamics. Beyond the transition the trajectory will spend some time in the region of the former first stable torus before it escapes to the region of the former second stable torus. The trajectory will chaotically jump between the two former stable tori. With increasing distance from the point of the appearance of the SNA at $\varepsilon = \varepsilon_c$ the mean time between two successive jumps $\langle\tau\rangle$ becomes shorter leading to a characteristic scaling behavior similar to crisis-induced intermittency [Sturman 1999a]: $\langle\tau\rangle \sim (\varepsilon - \varepsilon_c)^{-\beta}$. The scaling exponent $\beta = 1/2$ found by Sturman [1999a] coincides with the theory of crisis-induced intermittency [Sommerer et al. 1991].

6.4 Loss of transverse stability: blowout transition to SNA

Many physical systems possess certain symmetries and their dynamics includes therefore a symmetric invariant subspace S in state space. If a trajectory is initialized in S, then this trajectory will never leave S and the dynamics of the system will remain in this subspace for all times. Suppose that there exists an invariant set in the subspace S which is stable with respect to all perturbations in S. For general nonlinear dynamical systems this invariant set can be a fixed point, a periodic orbit, a torus or a chaotic set. Let us now consider a trajectory which starts in some distance from the invariant subspace S. Such a trajectory can be either attracted to S or repelled from S. The behavior of such a perturbed trajectory depends on the sign of the largest transverse Lyapunov exponent $\Lambda_{\perp}$ which determines the stability of the dynamics in the subspace S. This transverse Lyapunov exponent is computed for trajectories in S with respect to perturbations transverse to S. If $\Lambda_{\perp}$ is negative then all trajectories are attracted to S and the invariant set in S is an attractor. However, if the transverse Lyapunov exponent is positive, then the invariant set in S is transversely unstable and a new attractor appears which is not confined to the invariant subspace S anymore. It has also transverse components. This bifurcation is called *blowout bifurcation* and has been studied extensively in general chaotic systems possessing certain symmetries [Ott and Sommerer 1994; Ashwin et al. 1994; Lai and Grebogi 1995].

A similar transition can lead to the creation of a strange nonchaotic attractor [Kuznetsov et al. 1995; Yalcinkaya and Lai 1996, 1997] in quasiperiodically forced systems having appropriate symmetry properties. Suppose that the system possesses an attractor in an invariant subspace S. Upon varying the system parameters, this attractor can become transversely unstable, thus giving rise to the emergence of another attractor which extends then in a larger part of the state space containing the invariant subspace. In quasiperiodically forced systems we can distinguish different bifurcations which manifest themselves as blowout bifurcations. In Table 6.1 possible blowout bifurcations are listed: one of them leads to the formation of a strange nonchaotic attractor, the other three result in a chaotic attractor. Note that the attractor in the invariant subspace which becomes transversally unstable can be either a torus, a strange nonchaotic or a chaotic

Table 6.1 Possible blowout bifurcations

attractor in the invariant subspace prior to the bifurcation	blowout attractor beyond the bifurcation
torus	SNA
torus	chaos
SNA	chaos
chaos	chaos

attractor.

The simplest example to demonstrate the emergence of an SNA due to transverse instability is the famous GOPY-map (2.11, 2.12). This map possesses an invariant subspace S which is the line $(x = 0, \forall\theta)$. The whole subspace S is a quasiperiodic attractor for $a < 1$ (see Fig. 6.23a). According to Grebogi et al. [1984] this attractor loses its stability at the critical value $a = 1$. Let us look at the transverse Lyapunov exponent $\Lambda_\perp$. It can be computed analytically taking into account that we compute the derivative only for a trajectory in S, i.e. $x_n = 0, \forall n$

$$\Lambda_\perp = \left\langle \left| \ln \frac{dx_{n+1}}{dx_n} \right| \right\rangle = \left\langle \ln |(2a \cos 2\pi\theta_n)| \left. \frac{1}{\cosh^2 x_n} \right|_{x_n=0} \right\rangle = \ln |a| \quad (6.14)$$

The transverse Lyapunov exponent $\Lambda_\perp$ changes its sign as a crosses 1, while the usual Lyapunov exponent remains negative. As a result a strange nonchaotic attractor appears (see Fig. 6.23b). This SNA, which is not confined to the invariant subspace anymore, contains the former attractor as an unstable set giving rise to local expansion.

Besides the emergence of a strange nonchaotic attractor from a torus due to the loss of transverse stability as shown using the GOPY-map as an example, two other blowout-like bifurcations can occur leading to a chaotic attractor (see Table 6.1). Both these transitions we demonstrate using another paradigmatic example, namely two coupled logistic maps.

The study of coupled identical systems usually leads to a dynamics which contains invariant subspaces corresponding to a synchronized behavior of some of the subsystems. The synchronized attractor is then located in an invariant subspace, also called synchronization manifold. The loss of synchronization can be described in terms of bifurcations which cause the emergence of attractors involving parts of the state space outside the synchronization manifold. Blowout bifurcations are one possible mechanism of the loss of synchronization [Pikovsky et al. 2001]. In unforced maps they occur only when the attractor in the synchronization manifold is chaotic.

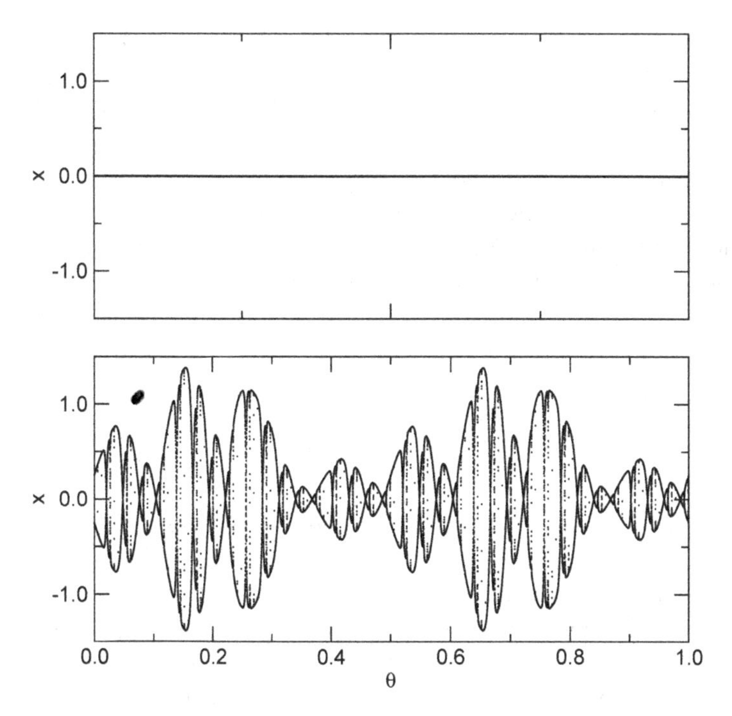

Fig. 6.23 The emergence of an SNA due to the loss of transverse stability in the GOPY-map: (top panel) stable torus in the invariant subspace S prior to the transition ($a = 0.9$); (bottom panel) SNA beyond the transition ($a = 1.1$).

This situation changes when a quasiperiodic forcing is introduced. Then we observe all the possible transitions listed in Table 6.1 as mechanisms for the loss of synchronization. It is important to note that in the forced case the dynamics in the invariant subspace prior to the bifurcation does not need to be chaotic.

To demonstrate this we consider two coupled logistic maps which are both forced quasiperiodically:

$$x_{n+1} = rx_n(1 - x_n) + b(y_n - x_n) + \varepsilon \cos(2\pi\theta_n) \,, \tag{6.15}$$

$$y_{n+1} = ry_n(1 - y_n) + b(x_n - y_n) + \varepsilon \cos(2\pi\theta_n) \,, \tag{6.16}$$

$$\theta_{n+1} = \theta_n + \omega \pmod 1 \,, \tag{6.17}$$

where r is the nonlinearity as usual, b denotes the coupling constant and ε the forcing amplitude. In general the dynamics takes place in the three-dimensional state space (x, y, θ). But synchronous motion of the two variables x and y can be observed in the invariant subspace

$S = \{(x, y, \theta) : x = y\}$ in certain regions of the parameter space spanned by r, b, ε. Thus the synchronized attractor lives in the diagonal plane S, and the dynamics is governed by a single forced logistic map (6.3, 6.2). According to our previous studies (cf. Chapter 2) the attractor in the synchronization manifold can be either a torus, a strange nonchaotic or a chaotic attractor. When a parameter is varied, this attractor in the invariant subspace loses its transverse stability and gives rise to a blown out attractor which extends also in the direction perpendicular to the synchronization manifold. To monitor the bifurcation one has to compute the transverse Lyapunov exponent which in this case is defined as

$$\Lambda_{\perp} = \lim_{N\to\infty} \frac{1}{N} \sum_{n=1}^{N} \ln |f'_x(x_n) - 2b| \,, \tag{6.18}$$

where f'_x denotes the derivative of the logistic function $f(x) = rx(1-x)$ with respect to x. Negative values of $\Lambda_{\perp}$ characterize regions in parameter space where the synchronized attractor is at least weakly stable in the Milnor sense [Milnor 1985; Pikovsky and Grassberger 1991]. Weak stability means for the synchronized attractor, that it is transversally stable in average, i.e. almost all trajectories on the attractor are transversally stable. For nonchaotic attractors on S, weak stability coincides with the usual Lyapunov stability. To characterize the dynamics within the invariant subspace, one uses the longitudinal Lyapunov exponent

$$\Lambda_{\parallel} = \lim_{N\to\infty} \frac{1}{N} \sum_{n=1}^{N} \ln |f'_x(x_n)| \,. \tag{6.19}$$

The overall dynamics can be identified by the usual Lyapunov exponent.

The blowout bifurcation leading from a torus to an SNA has been discussed above. Since the usual blowout bifurcation from chaos to chaos is studied in great detail in literature [Pikovsky et al. 2001] we focus here only on the transitions torus to chaos and SNA to chaos which are typical for quasiperiodically forced systems. Fig. 6.24 shows the transition from a synchronized motion on a torus directly to unsynchronized chaos. We show the attractor in two different projections, once in the (y, x) plane and once in (θ, x) plane to illustrate the character of the motion. Fig. 6.25 shows also the transverse $\Lambda_{\perp}$ and the usual maximal λ_{max} Lyapunov exponent. As the transverse Lyapunov exponent changes its sign, the chaotic blowout attractor is born.

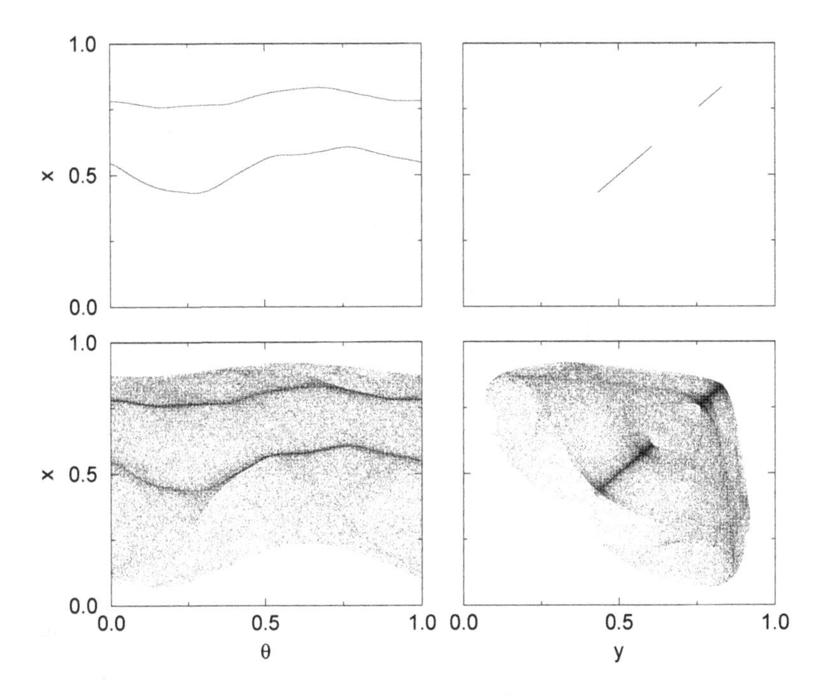

Fig. 6.24 Blowout bifurcation from synchronized torus T2 to a nonsynchronized chaotic attractor for $r = 3.22; \varepsilon = 0.03$ depending on the coupling constant b in different projections in state space: synchronized attractor for $b = 0.21$ (top panels); nonsynchronized attractor for $b = 0.23$ (bottom panels).

The same kind of transition (synchronized torus $\rightarrow$ unsychronized chaos) can also happen in a different form where the unsynchronized attractor first occupies only a small piece of the state space before it becomes larger in a second bifurcation (Figs. 6.26 and 6.27). This latter transition could also be interpreted in terms of an interior crisis (cf. Section 6.2.4) since we observe a sudden increase in the size of the attractor. The transition which resembles an interior crisis leads to a drastic increase in the value of the largest Lyapunov exponent as shown in Fig. 6.27.

Blowout bifurcations are in many cases related to the emergence of an extreme type of intermittency, the so-called on-off intermittency. To study the intermittent dynamics the distance of the trajectory to the synchronization manifold is investigated. This dynamics is characterized by laminar phases where this distance is close to zero, since the dynamics takes place close to the invariant subspace and turbulent bursts which correspond to large excursions of the trajectory away from the invariant subspace. The

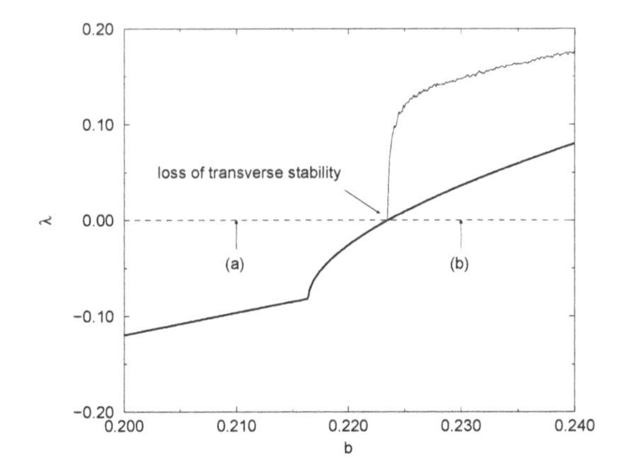

Fig. 6.25 Maximal (thin line) and transverse (bold line) Lyapunov exponent along the transition torus $\rightarrow$ chaos for $r = 3.22; \varepsilon = 0.03$. The arrows indicate the parameter values for which the attractor is shown in Fig. 6.24: (a) - Fig. 6.24 (top panels), (b) - Fig. 6.24 (bottom panels).

dynamics of the emerging unsynchronized attractor exhibits such on-off intermittency as depicted in Fig. 6.28. By contrast, the attractor arising from the interior crisis-like transition does not show on-off intermittency but crisis-induced intermittency.

The last kind of blowout bifurcation is illustrated in Figs. 6.29 and 6.30. In this case the synchronized attractor is an SNA while the unsynchronized attractor is chaotic.

It is important to note that the blowout bifurcation in quasiperiodically forced systems shares with the usual blowout bifurcation in chaotic systems the feature that the loss of transverse stability of the attractor in the invariant subspace causes the transition. Furthermore, chaotic systems showing a blowout bifurcation exhibit an extreme type of intermittency, the on-off intermittency [Platt et al. 1989; Pikovsky 1984; Fujisaka and Yamada 1985]. This type of intermittency can also be observed in quasiperiodically forced systems for certain blowout transitions [Yalcinkaya and Lai 1996]. However, other properties of the blowout bifurcation like its interpretation in terms of unstable periodic orbits fails in the case of quasiperiodically forced systems since the attractor in the invariant subspace is in most cases a regular attractor (torus or SNA) and not a chaotic one. But only chaotic attractors are covered by an infinite set of unstable periodic orbits. While these unstable periodic orbits and their transverse stability can be used to explain the mechanism of the blowout bifurcation in chaotic systems, this

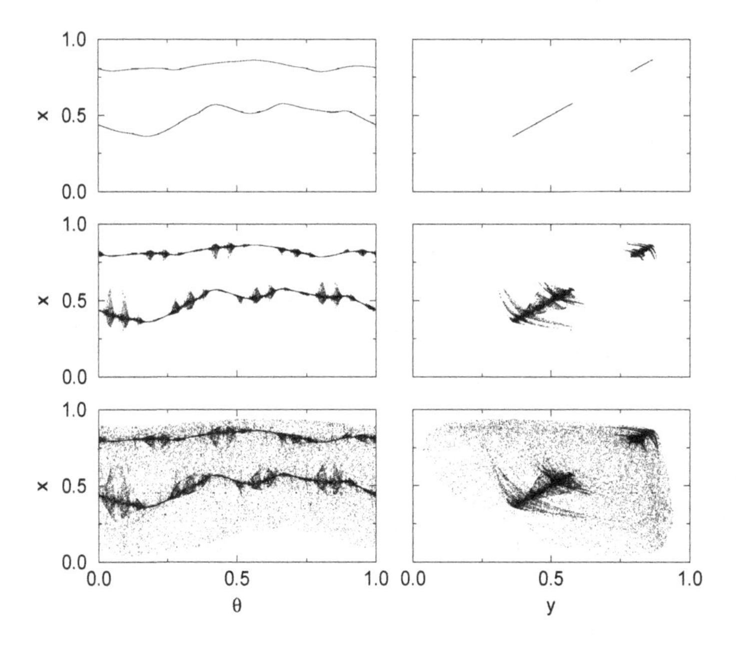

Fig. 6.26 Blowout bifurcation from synchronized torus T2 to a nonsynchronized chaotic attractor for $r = 3.35; \varepsilon = 0.03$ depending on the coupling constant b in different projections in state space: synchronized attractor for $b = 0.16$ (top panels); nonsynchronized attractor for $b = 0.18$ (middle panels); nonsynchronized attractor beyond interior crisis-like transition for $b = 0.19$ (bottom panels).

concept is not applicable to explain most of the transitions from Table 6.1 which occur in quasiperiodically forced systems.

6.5 Intermittency

Intermittency denotes a type of behavior where the dynamics varies chaotically between two different phases of motion. One of these phases is regular (close to stationary, periodic or quasiperiodic motion) and is called laminar phase. The laminar phase is interrupted by turbulent bursts which correspond to some irregular phase of motion. In chaotic systems three different types of intermittency are known which are related to three different inverse bifurcations: saddle-node (type I), Neimark-Sacker (torus) bifurcation (type II) and period-doubling (type III) [Pomeau and Manneville 1980a]. All these types of intermittency yield chaotic behavior when a system's parameter is varied. This is manifested by more and more frequent turbulent

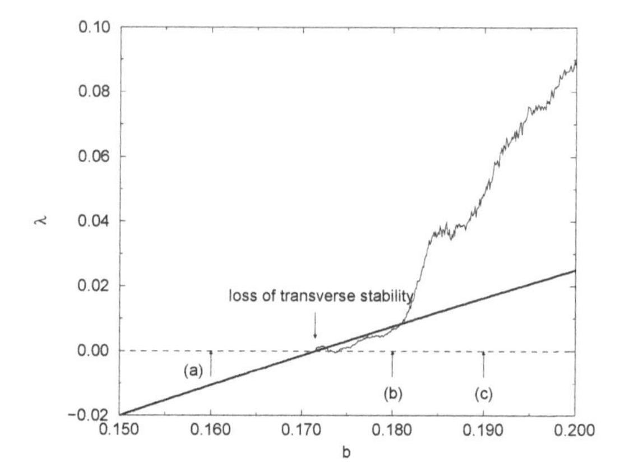

Fig. 6.27 The maximal (thin line)and the transverse (bold line) Lyapunov exponent along the transition torus → chaos for $r = 3.35; \varepsilon = 0.03$. The arrows indicate the parameter values for which the attractor is shown in Fig. 6.26: (a) - Fig. 6.26 (top panels), (b) - Fig. 6.26 (middle panels), (c) Fig. 6.26 (bottom panels).

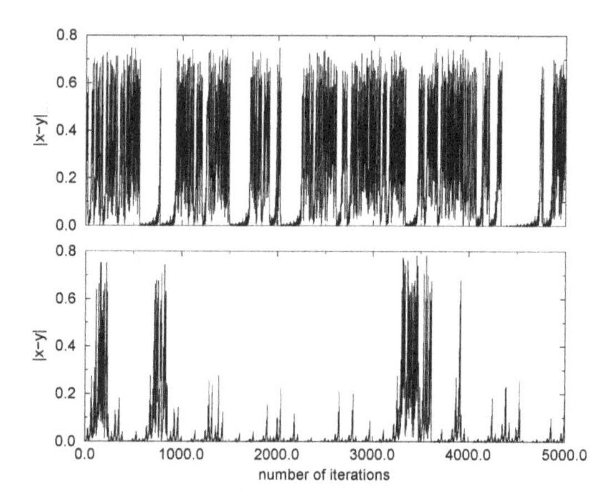

Fig. 6.28 Characteristics of the intermittent behavior for the two different transitions torus → chaos: (top panel) on-off intermittency for $r = 3.22$; (bottom panel) crisis-induced intermittency for $r = 3.35$ beyond the interior crisis-like transition.

bursts. The mean time between the appearance of bursts becomes shorter and changes according to certain scaling laws which are characteristic to the different types of intermittency.

In quasiperiodically forced systems intermittent behavior can also be related with the emergence of SNA [Prasad et al. 1997, 1998; Kuznetsov

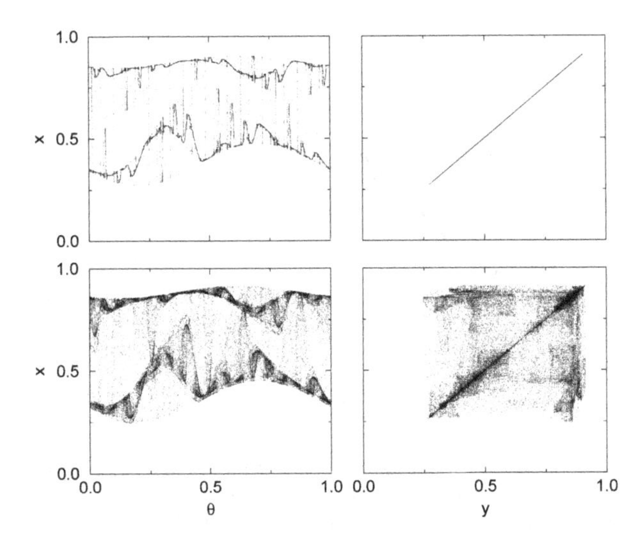

Fig. 6.29 Blowout bifurcation from synchronized SNA to a nonsynchronized chaotic attractor for $b = 0.01$; $\varepsilon = 0.03$ depending on the nonlinearity r in different projections in state space: synchronized attractor for $r = 3.51$ (top panels); nonsynchronized attractor for $r = 3.53$ (bottom panels).

2002a]. Two different intermittent routes leading to SNA have been found in quasiperiodically forced systems. One of them is related to the type I intermittency introduced by Pomeau and Manneville [1980a], the other one is connected with a sudden enlargement of the attractor and can be also interpreted in terms of an interior crisis (cf. Section 6.2.4).

Let us first discuss the generalization of the type-I intermittency to quasiperiodically forced systems. As already mentioned, type-I intermittency appears in general close to an inverse saddle-node bifurcation. In quasiperiodically forced systems such saddle-node bifurcations occur in two different forms: firstly we observe a smooth merging of a stable and an unstable torus leading to three-frequency motion on T^3 (see Section 6.3.1) for small forcing amplitudes, or, secondly, we obtain a non-smooth collision of stable and unstable tori leading to an SNA (see Section 6.3.2). For the intermittency type-I route, the transition for small forcing amplitudes is rather trivial and consists again in a smooth merging of the stable and unstable torus, only for larger forcing amplitudes a nontrivial transition is possible. It resembles the non-smooth saddle-node bifurcations leading to SNA.

For the purpose of generalization of intermittency type-I we introduce

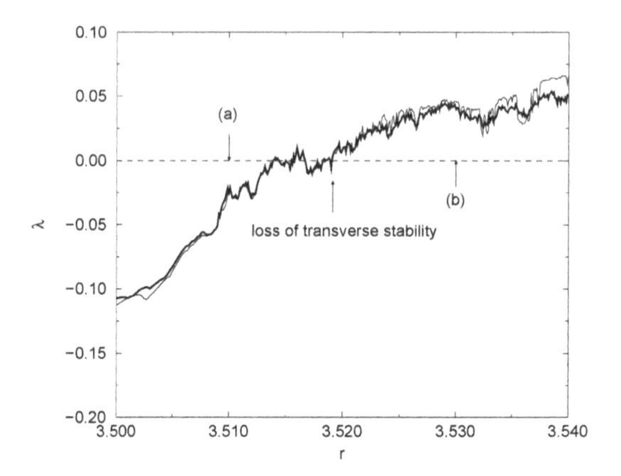

Fig. 6.30 The maximal (thin line) and the transverse (bold line) Lyapunov exponent along the transition SNA $\rightarrow$ chaos for $b = 0.01; \varepsilon = 0.03$. The arrows indicate the parameter values for which the attractor is shown in Fig. 6.29: (a) – Fig. 6.29(top panels), (b) – Fig. 6.29 (bottom panels).

a particular map which possesses the properties needed for an intermittent behavior [Kuznetsov 2002a]. As already mentioned above, intermittency is characterized by rather long laminar phases of the dynamics in the neighborhood of some former stable fixed point or periodic orbit, interrupted by turbulent bursts. After the bursts the dynamics comes close to the fixed point again. Hence, on one hand we need the presence of a saddle-node bifurcation, which ensures a dynamics corresponding to the laminar phase. Beyond the saddle-node bifurcation the trajectory travels through some small "channel" in the neighborhood of the former fixed points. On the other hand a certain re-injection mechanism is necessary which models the turbulent bursts and makes sure that the trajectory comes close the channel again to repeat the laminar phase. The following map fulfills theses conditions:

$$x_{n+1} = f(x_n) + b + \varepsilon \cos 2\pi\theta_n \ , \tag{6.20}$$

$$\theta_{n+1} = \theta_n + \omega \pmod 1 \ . \tag{6.21}$$

The function $f(x)$ is defined as follows:

$$f(x) = \begin{cases} x/(1-x) \text{ for } x \leq 0.75 \ , \\ 9/2x - 3 \text{ for } x > 0.75 \ . \end{cases} \tag{6.22}$$

One branch of this map is selected in form of a fractional-linear function $x/(1-x)$, which appears naturally in the analysis of the dynamics near a saddle-node bifurcation associated with intermittency. The other branch realizes the re-injection mechanism.

Without forcing ($\varepsilon = 0$) this map possesses the usual transition to chaos via type-I intermittency which is controlled by the parameter b. For $b < 0$ the map has two fixed points, one stable and one unstable, they merge at $b = 0$ and disappear. After that, for $b > 0$ a "narrow" channel remains at the former place of the two fixed points. There the trajectory has to travel across. The dynamics corresponding to the passage of the channel describes the laminar phase of motion. As $b \to 0$, the channel becomes narrower and narrower and the laminar phases become longer and longer. A detailed description of the intermittency transition can be found in standard textbooks on nonlinear dynamics and chaos.

When the forcing is applied, the transition happens for a pair of stable and unstable tori instead of the fixed points. Before the transition we observe in simulations a smooth torus while we obtain intermittent behavior beyond the transition (Fig. 6.31). As long as the forcing amplitude is small we still observe long laminar phases of motion close to the transition point. The pair of tori merges in each θ value as in a smooth saddle-node bifurcation, i.e. they coincide in the bifurcation point. However, with increasing forcing amplitude ε, the critical value of b where the tori merge is shifted from $b = 0$ to negative values. This merging can be described by a curve in the (ε, b) parameter space (cf. Fig. 7.4). As we increase the forcing amplitude ε further and move along the aforementioned bifurcation curve, we reach some critical value $\varepsilon = \varepsilon_c, b = b_c$, where the two tori still coincide in the transition point, but the tori have already a fractal-like structure. Beyond this critical point, the two tori only collide in a dense set of points like in a non-smooth saddle-node bifurcation. For our example the critical value is at $\varepsilon_c = 2, b_c = -0.597515...$ (cf. Section 7.3.4) which can be visualized by looking at the Lyapunov exponents as a function of b for ε fixed at the critical value $\varepsilon_c = 2$ (Fig. 6.32). The Lyapunov exponent is negative as long as $b < b_c$, but it becomes positive beyond the bifurcation point. The critical point which appears in this transition, the torus fractalization point (TF), will be discussed in detail once again in Chapter 7.

Let us now discuss the other intermittency transition leading to SNA which possesses several properties of an interior crisis. To describe the phenomenology of this transition we use the quasiperiodically forced quadratic map (6.1, 6.2) as a paradigm. The intermittent transition leading to an SNA can be observed in a small strip in the parameter space, which is located between quasiperiodic motion and chaos (see arrow F in Fig. 6.1).

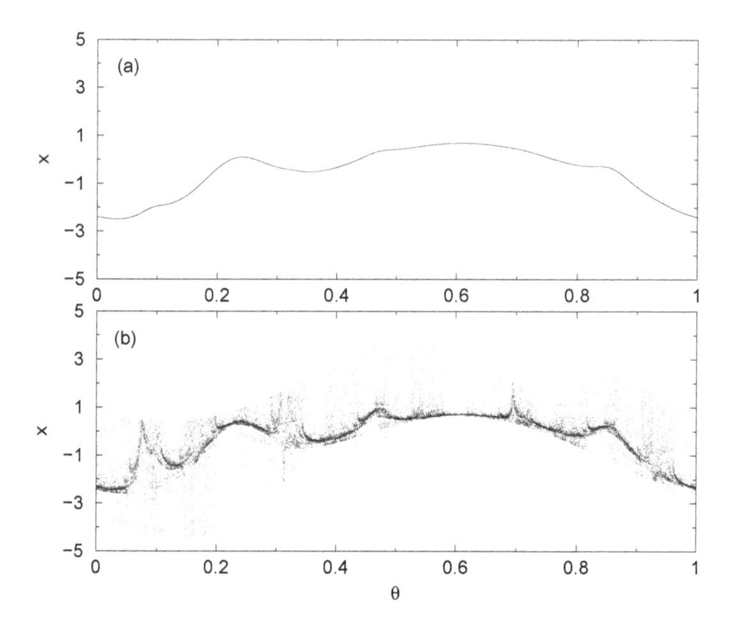

Fig. 6.31 Intermittency transition in the intermittency map (Eqs. 6.20, 6.21, 6.22) for $\varepsilon = \varepsilon_c = 2.0$ $b = -0.61$ (a) and $b = -0.59$ (b).

Suppose we consider the system in the parameter region where a stable torus exists (Fig. 6.33, top panel). Varying the forcing amplitude a sudden transition to an SNA is observed (Fig. 6.33, bottom panel). The dynamics on this SNA is intermittent, *i.e.* the trajectory spends most of the time in the neighborhood of the former torus and only seldom there are bursts which trace out a much larger fraction of the state space. Due to this dynamics this SNA has a special structure, where most of the natural measure occupies the former torus while only a few scattered points are found in a region far from the former attractor.

The mechanism of this transition has been pointed out and further analyzed by Kim et al. [2003b]. Using the method of rational approximations (cf. Chapter 3) they found that prior to the transition there exist several unstable saddles in the vicinity of the torus. At the bifurcation these saddles collide with the stable torus and finally they become part of the attractor. On one hand this yields a sudden enlargement of the attractor. On the other hand these saddles are responsible for the local expansion of nearby trajectories. Without computing the unstable invariant sets we can visualize the structure also by looking at successive rational approximations

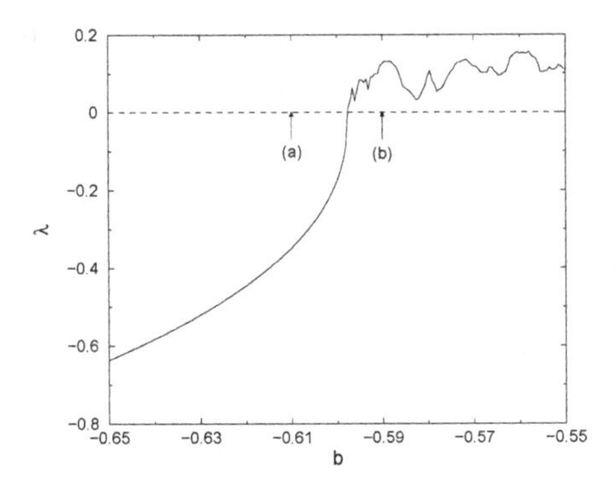

Fig. 6.32 The maximal Lyapunov exponent along the intermittency transition vs. b for $\varepsilon = \varepsilon_c = 2.0$. The arrows (a) and (b) indicate the parameter values of the phase portraits shown in Fig. 6.31.

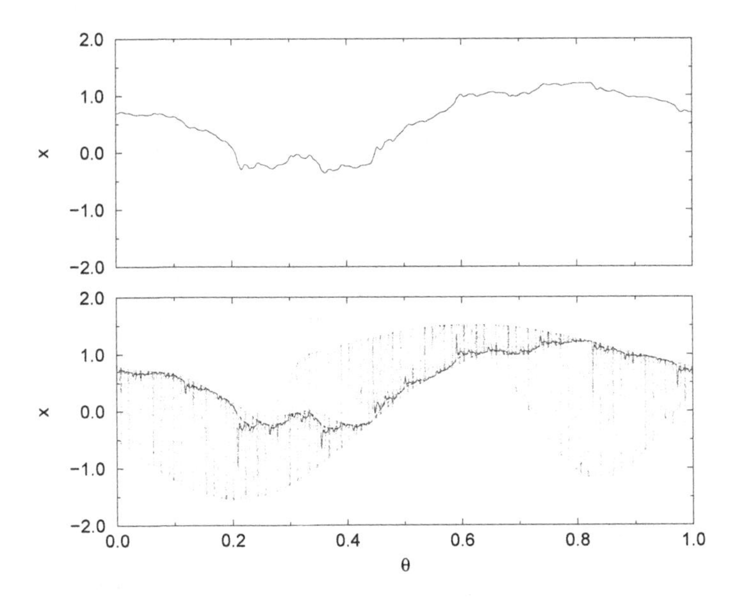

Fig. 6.33 Emergence of an intermittent SNA in the quasiperiodically forced quadratic map ($a = 1.1$). Top panel: torus attractor prior to the transition ($\varepsilon = 0.4193$); bottom panel: SNA beyond the transition ($\varepsilon = 0.4194$).

of the intermittent attractor (Fig. 6.34). Intermittency type-I transitions to chaos as well as interior crises are the dominant bifurcations which appear in rational approximations. They lead to sudden enlargements of the size of the approximate attractor for certain θ values. In the limit $k \to \infty$ these crises and intermittency type-I transitions yield the spiky structure of the attractor.

The intermittent SNA, which emerges in this transition is characterized by strong contraction in the neighborhood of the former stable torus and large expansions in the spiky regions. This complex behavior yields nontrivial distributions of local Lyapunov exponents (Fig. 6.35) with a pronounced maximum at positive values of the local exponents. The regions with chaotic dynamics which become embedded in the SNA are rather large in size. For this reason the existence intervals in the parameter space where these intermittent SNA appear are extremely small so that one can expect that they are very difficult to find in experiments. The transition to chaos happens very close to the transition to SNA.

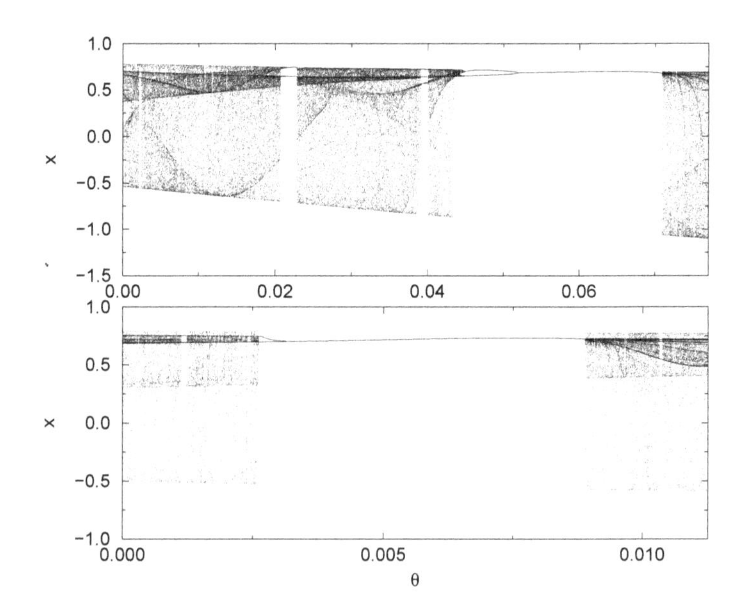

Fig. 6.34 Rational approximations for the emergence of an intermittent SNA in the quasiperiodically forced quadratic map ($a = 1.1, \varepsilon = 0.4194$). Top panel: $F_k = 13$; bottom panel: $F_k = 89$.

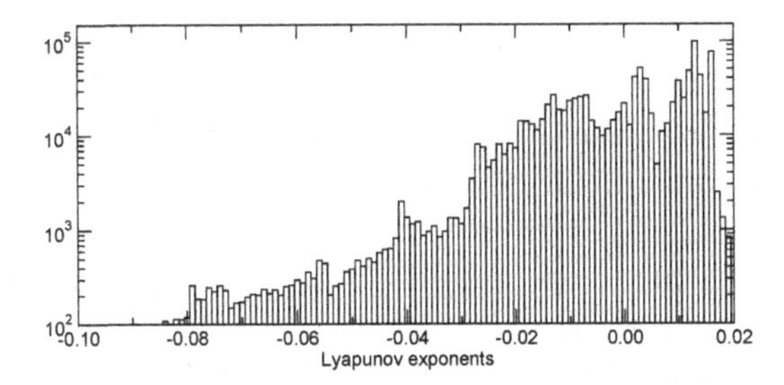

Fig. 6.35 Histogram for the distribution of finite time Lyapunov exponents for an intermittent SNA in the forced quadratic map ($a = 1.1; \varepsilon = 0.4194$).

6.6 Bibliographic notes

In this chapter we have focused on the transitions in quasiperiodically forced systems that include SNAs. Many authors have contributed to reveal the mechanisms of the bifurcations in quasiperiodically forced systems and to explain the emergence of SNAs. For mathematically rigorous results see [Broer et al. 1990]. A more practical approach to the calculation of bifurcations was presented by Chastell et al. [1995]. Algorithms to calculate rotation numbers in circle maps have been discussed in [Stark et al. 2002]. Torus doubling has been studied by Kaneko [1984c] and Stagliano et al. [1996].

Besides smooth bifurcations of tori corresponding to a merging of a stable and an unstable torus, non-smooth bifurcations are of particular importance in quasiperiodically forced systems because they are often related with the formation of SNAs. Non-smooth saddle-node bifurcations or tori collisions have been addressed as a mechanism of the emergence of SNA in the circle map by Feudel et al. [1995a]. The implications of this bifurcation on the shape of phase-locking regions are studied in [Ding et al. 1989b; Feudel et al. 1997; Glendinning and Wiersig 1999; Glendinning et al. 2000; Vasylenko et al. 2004]. Non-smooth tori collisions play also an important role in relation to other bifurcations: Heagy and Hammel [1994] discovered the appearance of SNA due to a collision of the doubled torus with its unstable parent, while Sturman [1999a] and Glendinning [2004] analyzed non-smooth pitchfork bifurcations. Such pitchfork bifurcations have been also studied related to the coexistence of several tori in resonance regions of the circle map [Osinga et al. 2001]. Non-smooth tori collisions and their consequences for the torus-doubling cascade have been addressed in [Heagy and Hammel 1994; Kuznetsov et al. 1995; Venkatesan and Lakshmanan

2001]. A collision of an invariant curve with itself is described in [Prasad et al. 1999].

The connection between smooth and non-smooth tori collisions has been studied using the renormalization group approach by Kuznetsov [2002b, 2003]. Special emphasis is given to the torus collision terminal point in [Kuznetsov et al. 2000], to the torus doubling terminal point in [Kuznetsov et al. 1995] and to the torus fractalization point in [Kuznetsov 2002a].

The fractalization of the torus is one of the first mechanisms of the appearance of SNA which has been discovered. The first to observe this phenomenon was Kaneko [1984b], who analyzed this problem later once again in detail in [Kaneko 1986; Nishikawa and Kaneko 1996]. Datta et al. [2004] explained the fractalization transition using rational approximations.

The blowout transition to SNA has been studied in the GOPY-map by Kuznetsov et al. [1995] and in other maps and flows by Yalcinkaya and Lai [1996, 1997]. In coupled systems the blowout transition has been studied related to the desynchronization of the dynamics in [Vadivasova et al. 2000; Neumann et al. 2003].

The intermittency route to SNA has been investigated by Kuznetsov [2002a] in terms of intermittency type-I according to the classification by Pomeau and Manneville [1980b] and by Venkatesan et al. [1999] in terms of a type-III intermittency. Another type of intermittency transition has been introduced by Prasad et al. [1997, 1998] in which the torus collides with an unstable chaotic set. This type of intermittency transition resembles an interior crisis because of the sudden widening of the attractor, the collision with the unstable chaotic set has been shown in [Kim et al. 2003b] using rational approximations. For other contributions concerning intermittency in quasiperiodically forced systems see [Glendinning 1998].

Several crises phenomena have been studied in quasiperiodically forced systems. The emergence of an SNA due to an interior crisis is observed in the quasiperiodically forced logistic map [Witt et al. 1997]. The boundary crisis has been analyzed in the quasiperiodically forced Hénon map by looking at the stable and unstable manifolds [Osinga and Feudel 2000] and in the logistic map by looking at rational approximations [Kim and Lim 2005]. As the last of the 3 different crises the band-merging crisis has been studied in the forced logistic map [Lim and Kim 2005].

Fractal basin boundaries for strange-nonchaotic attractors and their bifurcations due to quasiperiodic forcing have been considered in [Feudel et al. 1998b; Yang 2001; Shrimali et al. 2005].

Besides the usually used paradigmatic examples, bifurcations in higher-dimensional maps have been discussed in [Anishchenko et al. 1996; Sosnovtseva et al. 1996, 1998].

Most of the studies of bifurcations in quasiperiodically forced systems are devoted to maps. But there are also a few contributions to the investigation of flows [Lai et al. 1996; Venkatesan and Lakshmanan 1998; Venkatesan et al. 2000; Neumann and Pikovsky 2002; Kuznetsov and Neumann 2003].

Chapter 7

Renormalization group approach to the onset of SNA in maps with the golden-mean quasiperiodic driving

The renormalization group approach has been successfully applied to different types of the transition to chaos. In this chapter we demonstrate that it allows one to characterize the appearance of strange nonchaotic attractors. Because the methods used here are rather advanced, we start with a detailed introduction. Then, we define critical objects, for which it is possible to develop a renormalization group. In the description of the renormalization group analysis we dwell on the resulting scaling properties of regimes in the phase and the parameter space.

7.1 Introduction: The main idea of the renormalization group analysis

Originally, the approach called *renormalization group analysis* (RG) was developed in quantum field theory and in the theory of phase transitions [Shirkov et al. 1986; Balescu 1975]. Generally speaking, this is a tool to deal with objects possessing a wide interval of temporal and/or spatial characteristic scales.

In nonlinear dynamics the RG approach has been introduced by Feigenbaum [1979, 1983] and later has been successfully applied to the analysis of different types of transitions to chaos, e.g. via period doubling [Vul et al. 1984; Cvitanović 1989; Kuznetsov et al. 1997], intermittency [Hirsch et al. 1982; Hu and Rudnick 1982], and quasiperiodicity [Feigenbaum et al. 1982b; Rand et al. 1983]. As is commonly recognized, this is an effective and powerful theoretical instrument for uncovering of deep and fundamental features of the dynamics between order and chaos, in particular of quantitative universality and of scale invariance (scaling) for those subtle structures in phase space and in parameter space, which are associated with the transition.

To explain the general idea of the RG method in the context of nonlinear dynamics, let us assume that we have an evolution operator for some dynamical system on a particular time interval. Applying this operator several times, we construct an evolution operator for a larger time interval. An opportunity to use the RG analysis occurs in a rather specific, so-called *critical situation*, when it is possible to adjust parameters of the original system in such a way that the new operator for a larger time interval can be transformed exactly or approximately to the initial operator by a change of scales of the dynamical variable (variables). This procedure is called *an RG transformation*, and the adjusted parameters define the location of the *critical point* in the parameter space of the original system. The RG transformation may be applied repeatedly to obtain a sequence of evolution operators for larger and larger time scales.

A critical situation usually corresponds to convergence of the operator sequence to some definite limit, *a fixed point of the RG transformation*, or, as alternative, to *a periodic point* called also *a cycle*. However, the last possibility is not conceptually different, because in the case of period p one can speak of a fixed point of the RG transformation composed of p steps of the original construction.

The presence of a fixed point of the RG transformation means that the rescaled long-term evolution operators at the criticality will be of a universal form, up to a characteristic scale. In principle, this form of the renormalized operator may be recovered (say, numerically) directly from the operator fixed-point equation. The last is determined entirely by the structure of the RG scheme, i.e., without any reference to the originally examined system. Therefore, we assert that a fixed-point solution of a definite RG scheme gives rise to a *universality class*, which may include systems of very different mathematical nature (e.g. iterative maps, ordinary differential equations, extended systems, etc.)

What are the dynamical consequences of convergence to the fixed point of the rescaled operators produced by application of the RG procedure? Obviously, this convergence means that at the critical point the dynamics of the original system on different time scales are similar, up to the scale change of the dynamical variables. This is a property called *scaling*.

Moreover, scaling properties are intrinsic also to a vicinity of the critical point in parameter space. Let us suppose that we depart a little from the critical point, and the system demonstrates some kind of dynamical behavior. Then, we can decrease the displacement in such a way that the evolution operator after a larger number of steps of the RG transformation will be similar to that relating to the previous case. Therefore, we will have similar dynamics in the system, but with a larger characteristic time scale, and with smaller scales for the dynamical variables. As follows, a vicinity of the critical point in the parameter space contains a self-similar configuration of domains, which has the same structure for the entire universality class. (In fact, the scaling property appears as an asymptotic one: the smaller is a vicinity of the critical point, the higher is the accuracy of the observed self-similarity. On finite scales it may be regarded rather as approximation.)

As we speak about a vicinity of a critical point, we must examine small perturbations of the evolution operators near the fixed point of the RG transformation caused by detuning parameters from the critical situation. Under the assumption of smallness of the perturbations, it may be performed in terms of the linear stability analysis for this fixed point of the RG transformation. It gives rise to some eigenvalue problem. Relevant eigenvalues in the spectrum are those with the absolute value larger than 1 (that means, the perturbation grows under repetitive RG transformation). These eigenvalues play the role of *scaling factors* along appropriate axes in the parameter space. The number of relevant eigenvalues defines the *codimension* of the critical situation; this is the number of parameters to be adjusted to achieve criticality. For instance, in a three-dimensional parameter space critical situations of codimension one will occur at some surfaces, of codimension two at curves, and of codimension three at some points.

With respect to the solution of the RG equation, the relevant eigenvalues are associated with *an unstable manifold*, along which the orbits depart from the fixed point under application of the RG transformation. Irrelevant eigenvalues, which are less than 1 in modulus, correspond to *the stable manifold* and are responsible for the approach of the solutions to the fixed point, as the conditions of criticality are valid. It is clear from the present discussion, that for any type of critical behavior a fixed point of the RG transformation must be of *saddle type*.

As argued in the previous chapters, in quasiperiodically forced systems the occurrence of SNA is a typical attribute of the dynamics between order and chaos. Therefore, it is natural to expect that the RG analysis may be relevant for understanding the nature of SNAs, the mechanisms of their birth, and, perhaps, for fundamental quantitative regularities intrinsic to these phenomena. Originally, we advanced this idea in a note [Kuznetsov et al. 1995], devoted to the birth of SNA in the GOPY model (cf. Section 2.2.1) under quasiperiodic force (called later the "blowout" transition by Yalcinkaya and Lai [1996]). Then, we applied it to several types of critical points in parameter space of quasiperiodically driven maps (torus doubling terminal point [Kuznetsov et al. 1998], torus collision terminal point [Kuznetsov et al. 2000], and torus fractalization [Kuznetsov 2002a]). Up to now, only the case of the golden mean quasiperiodic forcing was studied in the frame of this approach. Here we present a review of these results and discuss possible directions of further developments.

7.2 The basic functional equations for the golden-mean renormalization scheme

Let us consider a concrete formulation of the RG approach, appropriate to the quasiperiodically forced dynamics with the golden-mean frequency ratio and start with the basic model of a forced one-dimensional discrete-time system

$$x_{n+1} = f(x_n, u_n) \, , \qquad u_{n+1} = u_n + \omega \pmod 1 \, , \tag{7.1}$$

where $\omega = (\sqrt{5} - 1)/2$. Note that contrary to our usual notation of the phase θ we denote it now as u. This change indicates that in the framework of renormalization theory this variable looses its specific importance as a phase but is understood as a usual variable describing a nonlinear process. Thus it undergoes all renormalization transformations in the same way as all other variables denoted by x.

The irrational value ω is approximated by ratios of subsequent Fibonacci numbers F_m, so, in the construction of the RG scheme it is natural to consider evolution operators for steps corresponding to the Fibonacci numbers. We write the evolution operator over F_m time steps as

$$x_{n+F_m} = f_m(x_n, u_n) \, , \qquad u_{n+F_m} = u_n + F_m \omega \pmod 1 \, , \tag{7.2}$$

in accordance with the definition of the Fibonacci numbers, $F_{m+2} = F_{m+1} + F_m$. Therefore, a subsequent application of the evolution operators over F_{m+1} steps and then over F_m steps yields

$$x_{n+F_{m+2}} = f_m(f_{m+1}(x_n, u_n), u_n + \omega F_{m+1}) \, . \tag{7.3}$$

To have a reasonable limit behavior of the sequence of evolution operators, we will change the scales for x and u by appropriate factors α and β at each new step of the construction. One of the relations associated with the golden mean reads

$$\omega F_{m+1} = -(-\omega)^{m+1} \pmod 1 . \tag{7.4}$$

As it follows from this equation, the second rescaling factor responsible for the scaling of u_n should be chosen as $\beta = -1/\omega = -1.618034...$. Now, instead of f_m, we introduce the renormalized functions

$$g_m(x, u) = \alpha^m f_m(x/\alpha^m,\ (-\omega)^m u) . \tag{7.5}$$

Rewriting (7.3) in terms of these functions we come to our RG equation [Kuznetsov et al. 1998, 2000; Kuznetsov 2002a, 2003]

$$g_{m+2}(x, u) = \alpha^2 g_m(\alpha^{-1} g_{m+1}(x/\alpha,\ -u\omega),\ \omega^2 u + \omega) . \tag{7.6}$$

In the particular case, when the functions g_m do not depend on the second argument, this functional equation reduces to the equation derived earlier to describe the critical behavior in an autonomous circle map at the golden-mean rotation number. It is associated with a fixed point solution studied by Feigenbaum et al. [1982b] and by Rand et al. [1983]. Our equation (7.6) is a two-dimensional generalization of that one. In the present Chapter, we shall deal with several different fixed points or cycles of this generalized equation. The constant α is specific for each of these solutions (universality classes), and has to be evaluated in the course of the numerical solution of the functional equation for each critical situation under study.

As explained in the introductory section, the next step in the RG analysis consists in the consideration of the dynamics in a vicinity of a fixed-point or cycle solution, i.e., in its stability analysis. For a cycle of period p, $g_1(x, u) \to g_2(x, u) \to ... \to g_p(x, u) \to g_1(x, u)$, we examine a perturbed solution $g_m(x, u) + \epsilon h_m(x, u)$, $\epsilon \ll 1$, and in the first order in ϵ we obtain the following equation:

$$\begin{aligned} h_{m+2}(x, u) = \alpha g_m'(\alpha^{-1} g_{m+1}(x/\alpha,\ -u\omega),\ \omega^2 u + \omega)\, h_{m+1}(x/\alpha,\ -u\omega) \\ +\alpha^2 h_m(\alpha^{-1} g_{m+1}(x/\alpha,\ -u\omega),\ \omega^2 u + \omega) . \end{aligned} \tag{7.7}$$

Together with the additional condition $h_{m+kp}(x, u) = h_m(x, u)\delta^k$, $m = 1, 2, ...p$, it defines an eigenproblem, the solution of which is the spectrum of δ values. In particular, for the fixed point case (period $p = 1$) the equation reduces to

$$\begin{aligned}\delta^2 h(x,u) = {} & \alpha\delta g'(\alpha^{-1}g(x/\alpha,\,-u\omega),\,\omega^2 u+\omega)h(x/\alpha,\,-uw) \\ & +\alpha^2 h(\alpha^{-1}g(x/\alpha,\,-u\omega),\,\omega^2 u+\omega)\,.\end{aligned} \tag{7.8}$$

Usually, the spectrum of eigenvalues requires a careful analysis for a correct interpretation.

First, we account only for eigenvalues, which are larger than 1 in modulus, because the relevant perturbation modes are those, which grow under iterations of the RG transformation.

Second, we must exclude eigenvalues associated with infinitesimal variable changes. For example, a shift $x \to x+\epsilon$, ϵ=const$\ll 1$, produces a variation of the original map, which may be interpreted as a perturbation of a solution of the RG equation. It grows by a factor α per step of the RG transformation. A shift $u \to u+\epsilon$ generates a perturbation, which grows by factor β. Obviously, such perturbations are not of interest because they may be removed by a trivial backward variable change.

Third, we must exclude eigenvectors, which correspond to a deviation of the solution of the RG equation from the so-called commutative subspace [Rand et al. 1983; Kuznetsov et al. 1998, 2000; Kuznetsov 2002a, 2003]. To explain this point, we remark that, in fact, there exists an alternative way to construct the RG transformation: we could produce the composition of F_m and F_{m+1} steps of iterations *in inverse order*, to obtain $x_{n+F_{m+2}} = f_{m+1}(f_m(x_n,\,u_n),u_n+\omega F_m)$. It leads to a different, although equivalent form of the whole theory. It is significant, however, that the terms of the functional sequence $f_m(x,u)$ must satisfy a condition of commutativity of the functional pairs, namely, $f_{m+1}(f_m(x_n,\,u_n),u_n+\omega F_m) = f_m(f_{m+1}(x_n,\,u_n),u_n+\omega F_{m+1})$, which implies also an analogous condition for the renormalized functions g_m. In the course of a formal solution of the eigenproblem (7.8), one gets among others some perturbation modes that violate this requirement, and they must be discarded.

Usually, the irrelevant modes can be recognized from the observation that their eigenvalues are some combinations of the factors α and β.

7.3 A review of critical points

Let us discuss briefly several critical situations, that can be analyzed in terms of the RG equation (7.6). As was already mentioned, the functions independent of the second argument represent a particular class of solutions of the functional equation. A solution of this kind is applicable to the critical point in the circle map discovered, by Shenker [1982] and studied in Refs. [Feigenbaum et al. 1982b; Rand et al. 1983] by means of the RG analysis. Other examples will correspond to solutions depending on both arguments and are associated with novel types of critical behavior in quasiperiodically driven maps: a map with pitchfork bifurcation, forced quadratic and circle maps, and forced fractional-linear map.

7.3.1 *Classic GM point*

The classical GM point theory has been developed in [Feigenbaum et al. 1982b; Rand et al. 1983; Shenker 1982]. On the parameter plane of the sine circle map (Fig. 7.1)

$$\varphi_{n+1} = \varphi_n + c + a \sin 2\pi\phi_n \quad (\mathrm{mod}\ 1)\ , \tag{7.9}$$

the critical line $a = a_c = 1/2\pi$ separates a region of regular (periodic or quasiperiodic) dynamics and an area where complex dynamics, including chaos, is possible. Below the critical line, periodic regimes occur inside the Arnold tongues, and quasiperiodic motions take place between them. They are associated with rational or irrational values of the rotation number, respectively, $\rho_\varphi(c,a) = \lim_{n\to\infty}(\varphi_n/n)$. There exists a curve of constant golden-mean rotation number: $\rho_\varphi(c,a) = \omega = (\sqrt{5}-1)/2$, which starts at $a = 0$, $c = \omega$, and meets the critical line $a = a_c = 1/2\pi$ at the GM critical point (GM stands for the 'golden mean'): $a = a_{GM} = 1/2\pi, c = c_{GM} = 0.60666106\ldots$

7.3.2 *Critical point of the blowout birth of SNA*

This section is based on the investigations in [Kuznetsov et al. 1995]. As mentioned, the first example of SNA has been found in the GOPY model with multiplicative quasiperiodic driving, and with a nonlinear function of the hyperbolic tangent [Grebogi et al. 1984]. For reasons of simplification of the subsequent RG analysis, we prefer to introduce a modified model

$$x_{n+1} = \frac{2\sigma x_n}{\sqrt{1+x_n^2}} \sin 2\pi n\omega\ , \qquad \omega = (\sqrt{5}-1)/2\ . \tag{7.10}$$

Qualitatively, the function $x/\sqrt{1+x^2}$ behaves in the same way as $\tanh x$. Therefore, all the arguments for occurrence of SNA remain valid for this map too. As follows, at $\sigma < 1$ the attractor is trivial, $x \equiv 0$, and at $\sigma > 1$

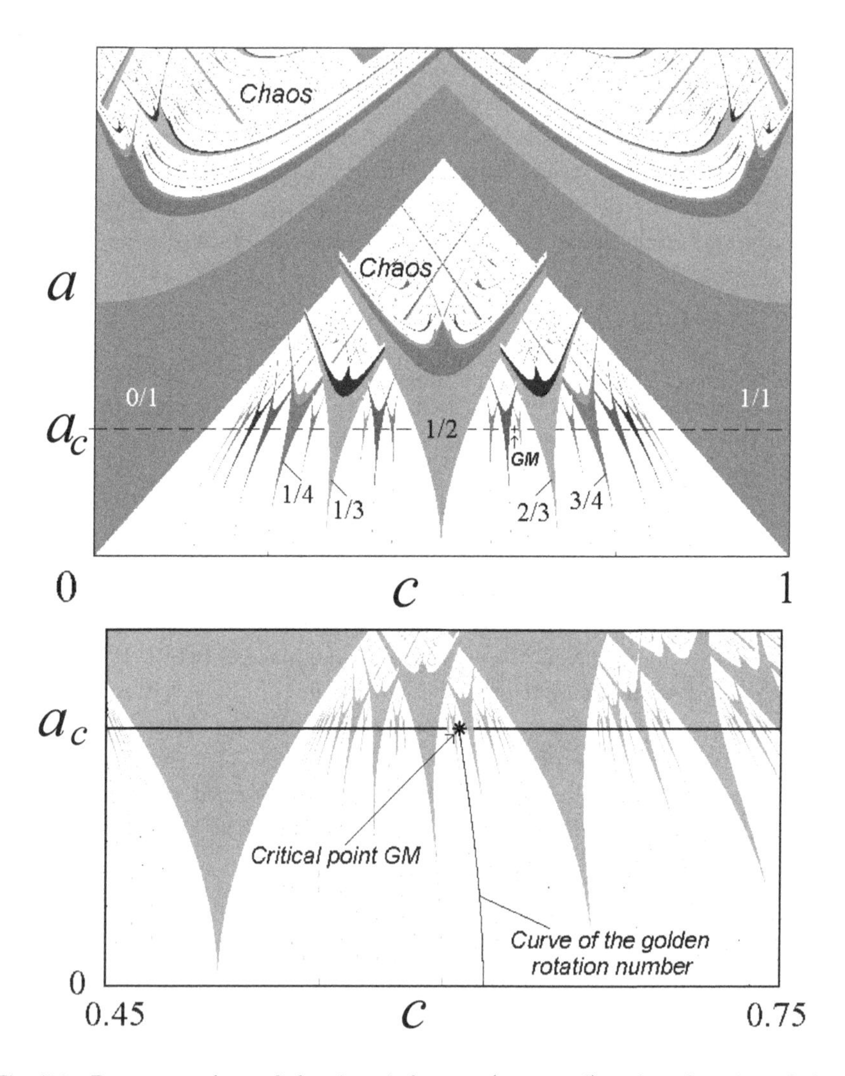

Fig. 7.1 Parameter plane of the sine circle map (top panel) and explanation of the location of the critical point GM (bottom panel). The fractions indicated inside the Arnold tongues are respective rotation numbers

it is an SNA. The transition called *the blowout birth of SNA* occurs at $\sigma = \sigma_{BO} = 1$, and this is precisely the critical point that may be studied in terms of RG analysis.

7.3.3 *Critical points of torus doubling terminal and torus collision terminal*

Here we follow the classification of [Kuznetsov et al. 1998, 2000]. The next model we will consider is a quasiperiodically driven logistic map [Kaneko 1984c; Kuznetsov 1984; Kuznetsov and Pikovsky 1989]

$$x_{n+1} = a - x_n^2 + \varepsilon \cos 2\pi n\omega \,. \tag{7.11}$$

Figure 7.2 shows a chart of dynamical regimes for this model on the parameter plane (ε, a). For $\varepsilon = 0$ Eq. (7.11) becomes a conventional logistic map. So, what is observed along the line $\varepsilon = 0$ is a usual period-doubling cascade, accumulating at the limit critical point of Feigenbaum (point F) [Feigenbaum 1979, 1983]. At small but nonzero ε, an increase of a gives rise to a sequence of torus doubling bifurcations discussed in Chapter 6. A stable fixed point is transformed into a stable smooth invariant curve, the torus-attractor T1. Instead of a stable period-2 orbit, we have an attractor consisting of two closed smooth curves, the doubled torus T2. A period-4 orbit, in turn, gives birth to a four-piece invariant curve (torus T4), and so forth. In contrast to the usual period-doubling cascade, the sequence of torus doublings is finite: the smaller the amplitude of driving, the larger is the number of the torus doubling bifurcations seen when increasing a [Kuznetsov et al. 1998, 2000; Kuznetsov 2002a, 2003; Shenker 1982; Grebogi et al. 1984; Kaneko 1984c]. If we keep a constant and increase the forcing amplitude, a smooth torus may transform into an SNA. In this transition, the Lyapunov exponent remains negative, but the geometrical structure of the attractor becomes complex, fractal-like. For larger a and ε, chaotic regimes with positive Lyapunov exponent arise. With further increase of the parameters, the orbits escape to infinity (white domain in Fig. 7.2).

On one side, the parameter interval $a \in (-0.25, 0.75)$ corresponding to the existence of an attractive fixed point in the unforced quadratic map is bounded by a period-doubling bifurcation, and on the other side by a tangent or saddle-node bifurcation associated with a collision of a pair of fixed points (stable and unstable) with their subsequent disappearance. Analogously, for the forced map, the top border of the domain T1 in Fig. 7.2 corresponds to the torus doubling bifurcation, and the bottom border to the tori collision: an attractor and an unstable set, represented by two invariant curves, approach each other, collide, and disappear.

Let us start at $\varepsilon = 0, a = 0.75$ and move on the parameter plane along the torus-doubling bifurcation curve. While the amplitude of driving is

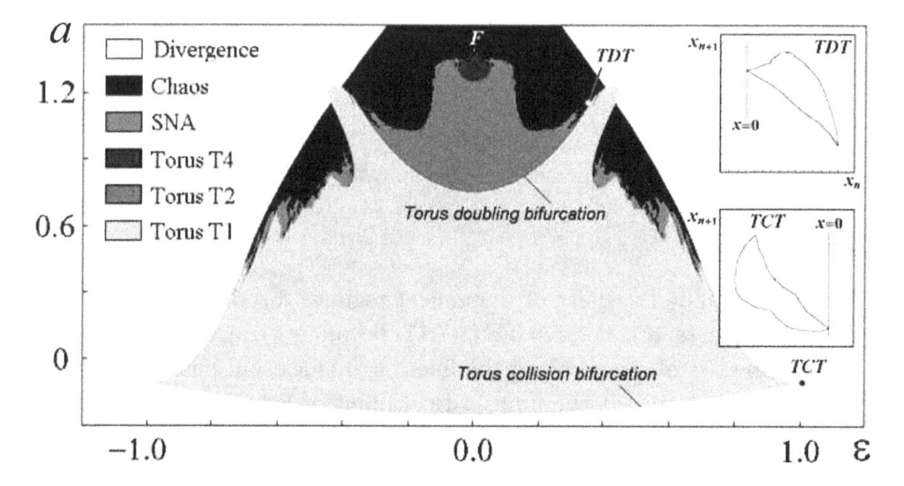

Fig. 7.2 Parameter plane of the quasiperiodically driven logistic map (7.11). Insets show invariant curves on iteration diagrams as they look at the TDT and TCT critical points. Legend is explained in the left top corner. F denotes the Feigenbaum accumulation point of the period doubling cascade.

small, the invariant curve born from the fixed point of the unforced system is of small width and placed on one (right) branch of the parabola. As the amplitude increases, the invariant curve grows in size and finally touches the location of the extremum $x = 0$ at some ε. At this moment, the torus doubling bifurcation line is terminated, and we call it the *TDT critical point* ('torus doubling terminal'). It occurs at $a = a_{TDT} = 1.15809685\ldots, \varepsilon = \varepsilon_{TDT} = 0.36024802\ldots$ [Kuznetsov et al. 1998].

Now, let us start at $\varepsilon = 0$, $a = -0.25$ and increase ε to move on the parameter plane along the torus collision bifurcation curve. The situation of collision of smooth invariant curves takes place while the invariant curve is confined on one (left) branch of the parabola. As the amplitude increases, the invariant curve grows in width and touches the location of the extremum $x = 0$ at some ε. This is the critical situation we call the *TCT critical point* ('torus collision terminal'). It occurs at $a = a_{TCT} = -0.09977123\ldots, \varepsilon = \varepsilon_{TCT} = 1.01105609\ldots$ [Kuznetsov et al. 2000].

Critical points of the same kind, TCT and TDT, were found also in the quasiperiodically forced sine circle map

$$x_{n+1} = x_n + c + a \sin 2\pi x_n + \varepsilon \cos 2\pi n\omega \pmod 1 \tag{7.12}$$

in the supercritical domain $a > 1/2\pi$, when the mapping near the extrema looks locally just like a parabola. In some respects, the sine circle map is a more convenient object for detailed study because no divergence occurs in this map as the variable x is defined modulo 1.

Figure 7.3 shows a chart of the dynamical regimes for the driven circle map on a part of the parameter plane (c, ε) including the TCT critical point [Kuznetsov et al. 2000]. The large gray domain in the diagram corresponds to the existence of the localized torus attractor. The right border of this domain is the bifurcation curve of the collision of a pair of smooth tori, one stable and another unstable one. Beyond the event, both of them disappear, and an intermittent chaotic regime arises, with a long-time travel through the region of the former existence of the attractor and the unstable invariant set (the 'channel'). Going along the bifurcation curve we see that the invariant curve at the situation of collision of the stable and unstable tori, grows in size, and ultimately touches the minimum of the map, there we arrive at the TCT point. As found numerically, it is located at $c = c_{TCT} = 0.377866239\ldots$, $\varepsilon = \varepsilon_{TCT} = 0.132566321\ldots$ The top border of the gray area corresponds to the situation of the intermittent transition from localized to delocalized SNA, with subsequent onset of chaos. The TCT critical point corresponds to the meeting of the bifurcation lines of the two mentioned distinct intermittent transitions.

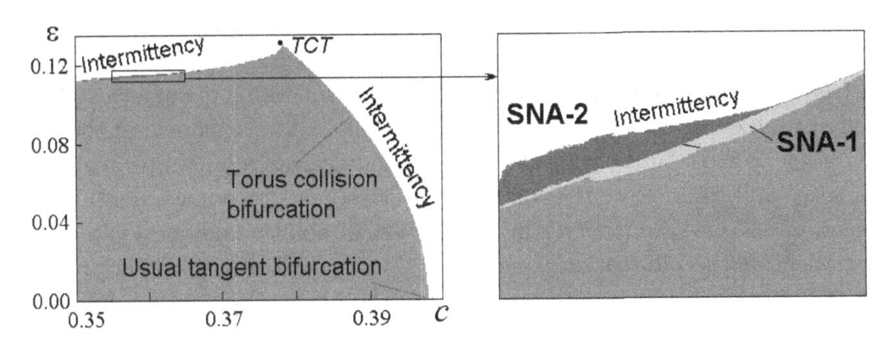

Fig. 7.3 Part of the parameter plane (c, ε) for the driven circle map (7.12) and an enlarged fragment that shows some subtle details near the intermittent transition.

7.3.4 *Critical point of torus fractalization*

As already mentioned, the fractalization of a torus and the transition to SNA in the forced sine circle map occurs in the critical and subcritical domains at $a \leq 1/2\pi$ [Kuznetsov 1984; Kuznetsov and Pikovsky 1989].

There, this transition cannot be associated with the TDT or TCT points because of the absence of a quadratic extremum. The nature of this criticality is associated with the *torus fractalization at the intermittency threshold* [Kuznetsov 2002a]. To describe this phenomenon we may use the model

$$x_{n+1} = f(x_n) + b + \varepsilon \cos 2\pi\omega n \,, \tag{7.13}$$

where $f(x)$ is defined as

$$f(x) = \begin{cases} x/(1-x), & x \leq 0.75 \,, \\ 9/2x - 3, & x > 0.75 \,. \end{cases} \tag{7.14}$$

One branch of the map is selected in a form of the fractional-linear function, $x/(1-x)$, which appears naturally in the analysis of the dynamics near the tangent bifurcation associated with intermittency (see, e.g., [Hirsch et al. 1982; Hu and Rudnick 1982]). The other branch is attached somewhat arbitrarily, to add the 're-injection mechanism' in the dynamics and to exclude divergence.

Figure 7.4 shows the chart of dynamical regimes for the model (7.13). The white area denotes a chaotic regime with positive Lyapunov exponent λ, and gray regions correspond to negative λ. In the bottom gray area, the attractor is a smooth torus. In the left part of the diagram the upper border of this region is a bifurcation curve of transition to a chaotic strange attractor via intermittency. The bifurcation is a merging of smooth stable and unstable tori with their coincidence. The Lyapunov exponent at the bifurcation is zero. In the right part, the bifurcation curve separates regimes of torus and SNA. Here the bifurcation corresponds to a non-smooth collision of two invariant curves in a dense set of points, and the Lyapunov exponent at the bifurcation is negative. The critical situation takes place at the point separating these two parts of the bifurcation border. We call it *the critical point of torus fractalization* (TF). In the model map (7.13) it is located at $\varepsilon_{TF} = 2, b_{TF} = -0.59751518\ldots$ [Kuznetsov 2002a].

7.4 RG analysis of the classic GM critical point

Critical behavior in the circle map associated with the break-up of the golden-mean quasiperiodicity (GM critical point) was studied in terms of RG analysis by Feigenbaum et al. and Ostlund et al. [Feigenbaum et al. 1982b; Rand et al. 1983]. In the frame of our general scheme, we can

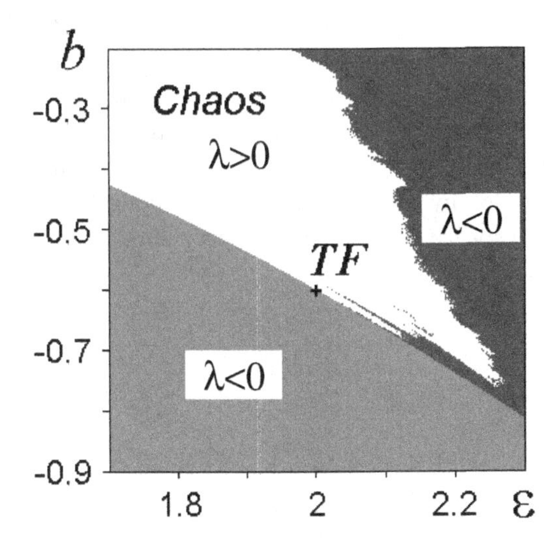

Fig. 7.4 Chart of dynamical regimes for the model (7.13). The bottom gray area corresponds to the existence of a smooth torus. The upper border is the bifurcation curve of the intermittent transition. In the left part the bifurcation corresponds to a merging of smooth stable and unstable tori with their coincidence, in the right part – to a non-smooth collision in a dense set of points. The white area denotes chaos, and dark gray presumably corresponds to SNA. The sign of the Lyapunov exponent λ is indicated in all three domains.

consider a set of two uncoupled maps

$$x_{n+1} = f(x_n), \qquad u_{n+1} = u_n + \omega \pmod 1 , \tag{7.15}$$

with $f(x) = x + c + a \sin 2\pi x$. The function f is independent of the argument u, thus the GM criticality will correspond to a degenerate fixed point of our functional equation (7.6): $g_k(x, u) \equiv G(x)$. In this case, Eq. (7.6) reads

$$G(x) = \alpha^2 G(\alpha^{-1} G(x/\alpha)) , \tag{7.16}$$

and it is known as the Feigenbaum–Kadanoff–Shenker equation. Numerically, the function was found in form of a high-precision expansion in powers of x^3 (see, e.g., [Feigenbaum et al. 1982b; Rand et al. 1983; Ivankov and Kuznetsov 2001]). The scaling constant is

$$\alpha = -1.288574553954 \ldots \tag{7.17}$$

Taking into account the representation of the sine circle map in the form (7.15), it is natural to depict the critical attractor in coordinates (u, x). As seen from Fig. 7.5, it looks like a fractal curve. Locally, the basic scaling property of this curve may be deduced from the RG analysis. Indeed, the evolution operators for time intervals increasing as Fibonacci numbers are asymptotically identical, up to the scale change. For each next Fibonacci number the variables x and u are rescaled by α and $\beta = -\omega^{-1}$. As follows, the attractor in coordinates (u, x) must possess self-similarity: increasing resolution in the plot by the factors α and β along the vertical and the horizontal axes, respectively, one can observe similar structures (see the bottom panels of Fig. 7.5).

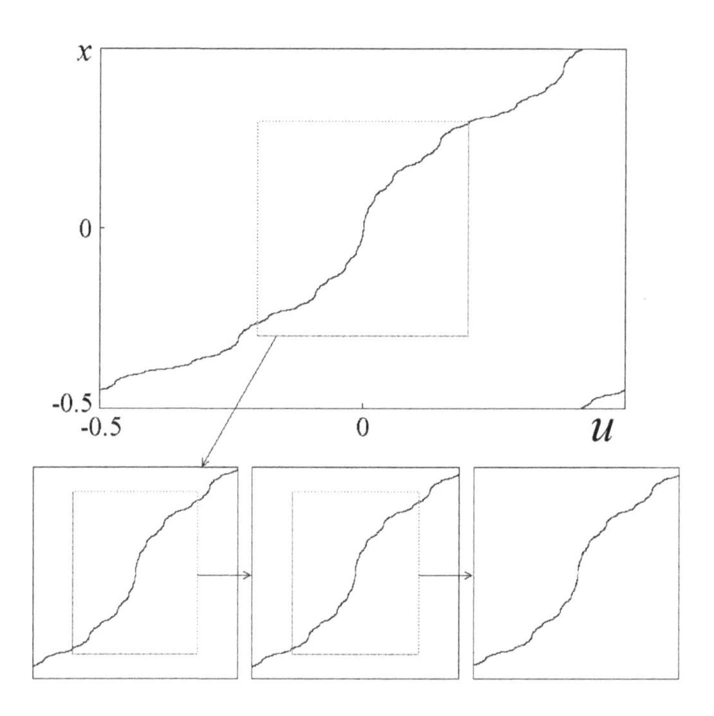

Fig. 7.5 Attractor of the map (7.15) at the GM critical point (top panel) and illustration of the basic local scaling property: the structure reproduces itself under magnification with factors $\alpha = -1.28857$ and $\beta = -1.61803$ along the vertical and the horizontal axes, respectively.

For perturbations of the GM fixed-point, Eq. (7.8) takes the form

$$\delta^2 h(x) = \alpha\delta G'(\alpha^{-1}G(x/\alpha))h(x/\alpha) + \alpha^2 h(\alpha^{-1}G(x/\alpha)) . \qquad (7.18)$$

The analysis of this equation (see, e.g., Refs. [Feigenbaum et al. 1982b; Rand et al. 1983; Ivankov and Kuznetsov 2001]) shows that there are two relevant eigenvalues,

$$\delta_1 = -2.8336106559\ldots, \qquad \delta_2 = \alpha^2 = 1.660424381\ldots. \tag{7.19}$$

These constants are responsible for the scaling properties of the parameter space structure near the GM critical point.

In general, to demonstrate a two-dimensional scaling, we have to define a special, perhaps curvilinear, local coordinate system near the critical point (*scaling coordinates*). The main requirement is that a displacement along each of the two coordinate axes has to give rise to a perturbation of the RG fixed point associated with one of the relevant eigenvalues (7.19).

In the case of the GM critical point it is natural to conclude that one coordinate line has to go along the critical line $a = 1/2\pi$, and another one along the curve of constant golden-mean rotation number. Numerically found expressions for parameters of the original map via new coordinates (c_1, c_2) are [Ivankov and Kuznetsov 2001]

$$c = c_c + c_1 - 0.01749c_2 - 0.00148(2\pi c_2)^2, \qquad a = a_c + 2\pi c_2\,. \tag{7.20}$$

In these equations we retain terms up to the second order because of the relation between δ_1 and δ_2: $\delta_2 < \delta_1$ and $\delta_2 < \delta_1^2$, but $\delta_2 > \delta_1^3$ (see Refs. [Kuznetsov et al. 1997; Ivankov and Kuznetsov 2001] for an explanation of the rules for the selection of the scaling coordinates). Figure 7.6 shows a chart of dynamical regimes with Arnold tongues and a sequence of fragments for several steps of magnification in the scaling coordinates. One can observe an excellent reproduction of the two-dimensional arrangement of the tongues at subsequent levels of resolution.

7.5 RG analysis of the blowout birth of SNA

To study the blowout bifurcation for the birth of SNA we have introduced a modified model with multiplicative quasiperiodic driving

$$x_{n+1} = \frac{2\sigma x_n}{\sqrt{1+x_n^2}}\sin 2\pi u_n\,, \quad u_{n+1} = u_n + \omega\,(\mathrm{mod}\ 1)\,, \quad \omega = (\sqrt{5}-1)/2\,. \tag{7.21}$$

This modified model exhibits the same transition to SNA like the GOPY model at $\sigma = 1$. This is a critical point, which may be studied in terms of our general RG scheme.

A reason for our preference of the nonlinear function used in (7.21) is that a composition of such functions generates again a function of the same

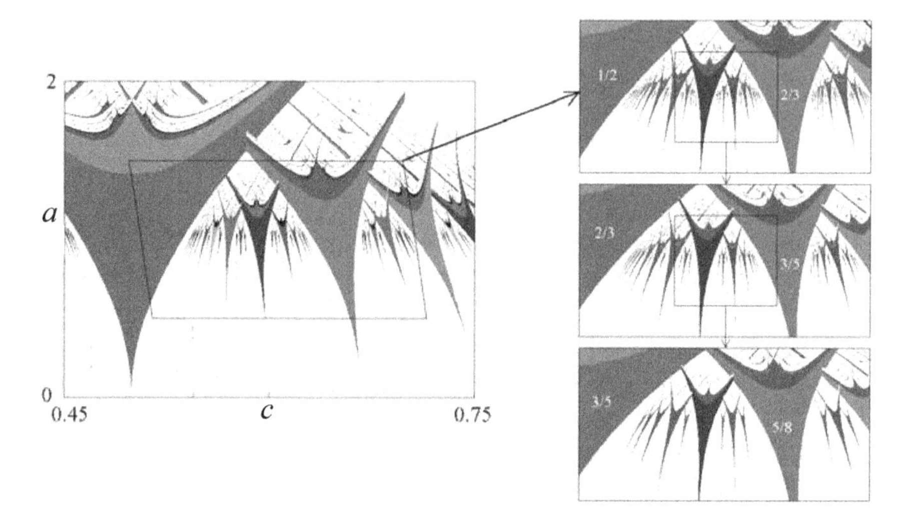

Fig. 7.6 A general picture of the parameter plane for the sine circle map and a sequence of fragments for several steps of magnification of a vicinity of the GM critical point in the scaling coordinates, with factors δ_1 and δ_2 along horizontal and vertical axes, respectively.

class. Namely, if $f_{1,2}(x) = P_{1,2}x/\sqrt{1+S_{1,2}x^2}$, then $f_1(f_2(x)) = f_3(x) = P_3x/\sqrt{1+S_3x^2}$, where $P_3 = P_1P_2$, $S_3 = S_2 + S_1P_2^2$. This is an important point in the framework of the RG analysis, because for an arbitrary number of iteration steps the evolution of the x variable will be given by a function of this special class. In particular, it relates to Fibonacci numbers of steps, which are of main interest as we deal with the golden mean frequency ratio.

Given a Fibonacci number F_m, we assume that $x_{n+F_m} = f_m(x_n, u_n)$, with $f_m(x,u) = P_m(u)x/\sqrt{1+S_m(u)x^2}$. A subsequent performance of F_{m+1} and then F_m iterations results in F_{m+2} steps, and we must have

$$f_{m+2}(x,\, u) = f_m(f_{m+1}(x,\, u), u+F_{m+1}\omega) = f_m(f_{m+1}(x,\, u), u-(-\omega)^{m+1})\,, \tag{7.22}$$

or, using the composition rules for the considered class of functions,

$$P_{m+2}(u) = P_{m+1}(u)P_m(u-(-\omega)^{m+1})\,, \tag{7.23}$$

$$S_{m+2}(u) = S_{m+1}(u) + S_m(u-(-\omega)^{m+1})\left[P_{m+1}(u)\right]^2\,. \tag{7.24}$$

Following our standard construction of the RG scheme, let us change scales of the dynamical variables, $x \to x/\alpha^m$, $u \to (-\omega)^m u$, where α is a rescaling factor to be determined later, and introduce the renormalized functions

$$g_m(x,u) = \alpha^m f_m(x/\alpha^m,\, (-\omega)^m u)\,. \tag{7.25}$$

Here

$$g_m(x,u) = Q_m(u)x/\sqrt{1+H_m(u)x^2}\,, \tag{7.26}$$

$$Q_m(u) = P_m((-\omega)^m u)\,, \tag{7.27}$$

$$H_m(u) = \alpha^{2m} S_m((-\omega)^m u)\,. \tag{7.28}$$

As follows from (7.22), (7.25), the functional sequence $g_m(x,u)$ will satisfy our general recurrent RG equation (7.6), and in terms of the coefficients Q_m, H_m it implies

$$Q_{m+2}(u) = Q_{m+1}(-\omega u)Q_m(\omega^2 u + \omega)\,, \tag{7.29}$$

$$H_{m+2}(u) = \alpha H_{m+1}(-\omega u) + \alpha^2\left[Q_{m+1}(-\omega u)\right]^2 H_m(\omega^2 u + \omega)\,. \tag{7.30}$$

To perform the functional iterations, it is sufficient to have the involved functions defined on the fundamental interval $u \in [-\omega, 1]$. Indeed, if u belongs to this interval, the same is true for the arguments in the right-hand parts of the equations, $(-\omega u)$ and $(\omega^2 u + \omega)$.

Note that the first equation (7.29) is independent of the second one (7.30). Hence, we can decompose the analysis onto two stages: first, we reveal the nature of the relevant solution for Q_m, and then examine the second equation, which is linear in H_m. It possesses squares $Q_m^2(u)$ as coefficients, so it is convenient to deal directly with them. Obviously, they obey an equation of the same form as (7.29):

$$Q^2_{m+2}(u) = Q^2_{m+1}(-\omega u)Q^2_m(\omega^2 u + \omega)\,. \tag{7.31}$$

To formulate the initial conditions we use that $F_1 = 1$ and $F_2 = 1$, set $P_1(u) = P_2(u) = 2\sin 2\pi u$ (that corresponds to the critical point of the model (7.21)), and define the rescaled functions, which relate both to *one* iteration step:

$$Q_1^2(u) = 2\sin^2(-2\pi\omega u),\ Q_2^2(u) = 2\sin^2(2\pi\omega^2 u)\,. \tag{7.32}$$

To explain what happens under iterations of the RG equation, we follow an elegant scheme developed by Mestel, Osbaldestin, and Winn [2000], although these authors deal with another solution of Eq. (7.29) in a different context than we do. The main idea is to consider the step-by-step generation of zeros of the functions at subsequent levels of construction. The functions under consideration are the so-called entire functions, which are determined completely by the distribution of their zeros.

One can see from Eq.(7.31) that the presence of a "mother" zero $z^{(m)}$ of a function $Q_m^2(u)$ (at the m-th level) implies the appearance of "daughter" zeros in functions $Q_{m+1}^2(u)$ and $Q_{m+2}^2(u)$ relating to the $m+1$st and $m+2$nd levels, respectively:

$$z^{(m)} \longrightarrow \begin{cases} z^{(m+1)} = -z^{(m)}/\omega \,, \\ z^{(m+2)} = (z^{(m)} - \omega)/\omega^2 \,. \end{cases} \tag{7.33}$$

To start, we consider zeros from a fundamental interval $[-\omega, 1]$, two in the function $Q_1^2(u)$ ($z_1^{(1)} = 0$ and $z_2^{(1)} = 1/2\omega$) and one in the function $Q_2^2(u)$ ($z_1^{(2)} = 0$). Then, from level to level, they are multiplying, precisely like Fibonacci's rabbits. Note that all zeros we deal with are of order 2 (double degenerate). Selecting some finite interval on the real axis, we can observe that at sufficiently high levels, the generated distributions of zeros manifest repetition after each third step, in other words, the sets of zeros are identical at m and $m+3$.

It means that the sequence of functions $Q_m^2(u)$ corresponds asymptotically to a period-3 cycle of the RG equation (7.31). The functions constituting this cycle are products of factors $(u - z_i^{(m+3k)})^2$ over an infinite set of zeros generated in the limit $k \to \infty$, with appropriately chosen normalization. The final solution of (7.31) is

$$Q_m^2(u) = \frac{1}{\omega} \lim_{L\to\infty} \lim_{k\to\infty} \prod_{|z|<L} \frac{(z_i^{(m+3k)} - u)^2}{|(z_i^{(m+3k)} + \omega)(z_i^{(m+3k)} - \omega)|} \,. \tag{7.34}$$

(The correct normalization is ensured by the expression in the denominator, which does not depend on u. It results from the fact that the periodic solution of (7.31) must satisfy $Q_m^2(\omega)Q_m^2(-\omega) = \omega^{-2}$, which may be derived from some manipulations with Eq.(7.31) taking into account that $Q_m(0) = 0$.)

Equation (7.30) for H_m is linear with periodic (in renormalization time m) coefficients, the period equals 3. So, the asymptotic solution is expected to have a form $H_{3(k+1)+m}(u) = \nu^k H_{3k+m}(u)$, where ν is some constant. Recall that the equation includes a rescaling factor α, yet not determined. We require it to be selected to ensure periodicity of the sequence H_m (i.e., to have $\nu = 1$), and under this condition we obtain from (7.30)

$$\begin{aligned} \alpha^{-2}H_1(u) &= \alpha^{-1}H_3(-\omega u) + Q_3^2(-\omega u)H_2(\omega^2 u + \omega)\,, \\ \alpha^{-2}H_2(u) &= \alpha^{-1}H_1(-\omega u) + Q_1^2(-\omega u)H_3(\omega^2 u + \omega)\,, \\ \alpha^{-2}H_3(u) &= \alpha^{-1}H_2(-\omega u) + Q_2^2(-\omega u)H_1(\omega^2 u + \omega)\,. \end{aligned} \tag{7.35}$$

This set of equations poses an eigenproblem, where α appears as an eigenvalue, and $H_{1,2,3}(u)$ are components of the eigenvector. To solve it numerically, we represent the functions $H(u)$ by tables of their values at nodes of a finite grid on the fundamental interval $[-\omega, 1]$, and use an interpolation scheme between the nodes. This reduces the set of functional equations to a finite set of algebraic equations. Then, the eigenproblem for a finite matrix may be solved by standard methods of linear algebra. The resulting eigenvalue equals

$$\alpha = 1.09804\ldots\,, \tag{7.36}$$

and this is the rescaling factor for the x variable per one step of the RG transformation.

If we wish to determine the functions $g_m(x,u)$, we need $Q_m(u)$ rather than $Q_m^2(u)$. It appears that with respect to index m these functions manifest repetition with period 6, which is twice larger, due to sign changes. As may be shown, they obey the relations

$$Q_1(u) = Q_4(u)\,, \qquad Q_2(u) = -Q_5(u)\,, \qquad Q_3(u) = -Q_6(u)\,, \tag{7.37}$$

and then we have

$$g_m(x,u) = Q_m(u)\frac{x}{\sqrt{1 + H_m(u)x^2}}\,,\ m = 1, 2, \ldots 6\,. \tag{7.38}$$

These functions constitute a period 6 cycle of the RG equation as illustrated in Fig. 7.7. The meaning of $g_m(x,u)$ is that it determines a rescaled evolution operator for F_{m+6k} iterations of the original map at the critical point in the limit $k \to \infty$. The cyclic repetition of the functions implies the presence of scaling regularities, i.e. of similarity of the dynamics on different time scales. However, formulating these scaling properties, we observe that the dynamics is similar to the original one not after six, but after three steps of the RG transformation: although the functions $g_m(x,u)$ and $g_{m+3}(x,u)$ can differ in sign, this is not essential because the model under study is symmetric in respect to a change $x \to -x$.

Three steps of the RG procedure correspond to the increase of the time scale by factor $\tau_m = F_{m+3}/F_m$, asymptotically $\tau = 1/\omega^3 = 4.236068$. Now,

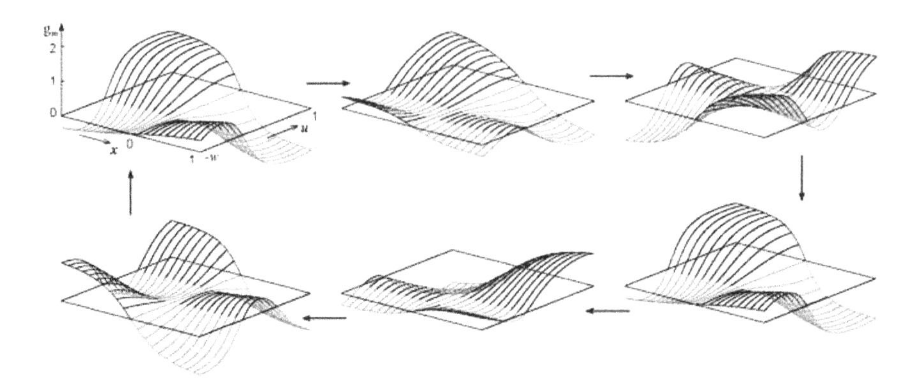

Fig. 7.7 3D plots of the functions $g_m(x, u)$ constituting the period-6 cycle of the RG equation.

if we observe a dynamical evolution starting from some x and u, then we get similar dynamics with time scale multiplied by τ, if we start from x/A and u/B, where $A = \alpha^3 = 1.32390$ and $B = 1/(-\omega)^3 = -4.236068$ are the phase space scaling factors.

The only observable kind of dynamics precisely at the critical point is a transient to the final trivial stationary state $x = 0$. From the above statements we conclude that this decay must follow a power law, $|x_n| \propto n^{-\gamma}, \gamma = \log A / \log \tau = 0.19436$.

These numbers are valid in a vicinity of $u = 0$, at which the scale transformation was defined in our derivation of the RG equation. At other initial phases, the scalings may be different (this reflects a multifractal nature of the SNA that is born in the transition under study). Particularly, with one iteration of the original map, a point (x, u) from a vicinity of $u = 0$ is mapped to a neighborhood of $u = \omega$, with a multiplication of x by $\sin 2\pi u \propto u$. Thus, the scaling factor for x near this new point will be a product of factors for x and u at the old point, namely, $A' = AB = \alpha^3/\omega^3 = 5.60809$. Respectively, the decay of x near $u = \omega$ will follow a distinct power law $|x_n| \propto n^{-\gamma'}, \gamma' = \log A' / \log \tau = \gamma + 1 = 1.19436$.

Figure 7.8 illustrates the scaling regularities intrinsic to the critical decay. Absolute values of x, generated by the map (7.21) at $\sigma = 1$, are plotted versus discrete time n in double logarithmic scale. Observe that the region occupied by the plotted points is bounded from above and from below by straight lines corresponding to the power laws $|x_n| \propto n^{-\gamma}$ and $|x_n| \propto n^{-\gamma'}$. Evidently, γ and γ' are just the largest and the smallest numbers among the critical indices associated with the decay at the threshold of birth of

SNA.

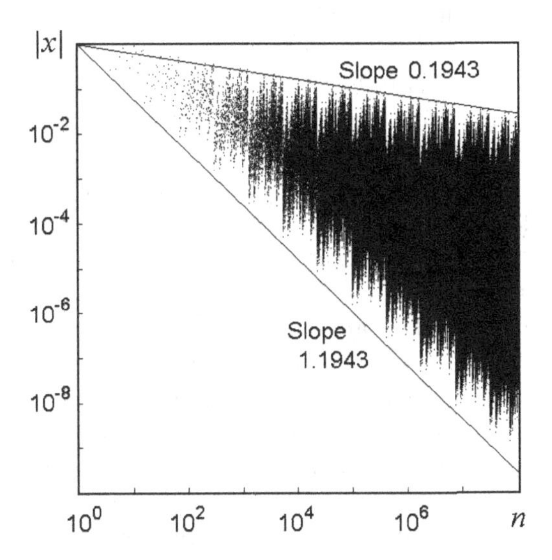

Fig. 7.8 Absolute value of the dynamical variable x versus discrete time at the critical parameter value $\sigma = 1$ plotted in double logarithmic scale. Initial value $x = 1$. Note that the observed values of x are bounded from above and from below by straight lines corresponding to the power laws $|x| \propto n^{\gamma}$ and $|x| \propto n^{\gamma'}$, respectively.

The second part of the RG analysis consists in the consideration of a perturbation for the periodic solution that describes a deviation from the critical point. Let us search for a solution of equation (7.29) as $Q_m(u) + \epsilon \tilde{Q}_m(u)$, $\epsilon \ll 1$. In the first order in ϵ we get

$$\tilde{Q}_{m+2}(u) = \tilde{Q}_{m+1}(-\omega u)Q_m(\omega^2 u + \omega) + Q_{m+1}(-\omega u)\tilde{Q}_m(\omega^2 u + \omega) \,. \quad (7.39)$$

In terms of our RG formalism, a shift of parameter $\sigma \to 1 + \Delta\sigma$ in the model map (7.21) corresponds to a multiplication of $Q_1(u)$ and $Q_2(u)$ by $1+\Delta\sigma$. As may be checked by direct substitution, a solution with initial conditions $\tilde{Q}_{1,2}(u) = Q_{1,2}(u)\Delta\sigma$ is $\tilde{Q}_m(u) = Q_m(u)F_m\Delta\sigma$. In the course of iterations of the RG equation, the same function (up to a sign) appears again after three steps, but multiplied by factor F_{m+3}/F_m. Asymptotically, this ratio yields a constant for renormalization of the parameter deflection

$$\delta = 1/\omega^3 = 4.236068 \,. \quad (7.40)$$

(We may not consider the second equation for $H(u)$, as the present argumentation is sufficient to get δ.)

Now we formulate a scaling property accounting for the parameter shift with respect to sustained dynamical regimes. Let us suppose that at a parameter value $\sigma = 1 + \Delta\sigma$ close to the critical point we observe some dynamical regime, say, SNA. Now, decreasing $\Delta\sigma$ by factor $\delta = 1/\omega^3$, we come to a situation, where the evolution operator is similar to the original one, but with a time scale increased by $\tau = 1/\omega^3 = 4.236068$, the scale in x decreased by $A = \alpha^3 = 1.32390$, and the scale in u reduced by $B = 1/(-\omega)^3 = -4.236068$. Drawn in rescaled coordinates, the SNA will look like the original one. (In fact, the scaling property is asymptotical: the closer σ to the critical point, the more accurate is the similarity.)

Figure 7.9 illustrates the stated scaling property of attractors in a vicinity of the critical point. The computed portraits of SNAs are plotted: dynamical variable x versus phase variable u. Each next picture corresponds to smaller distance from the critical point, decreased by factor δ, and is shown with magnification by factors a and b along the vertical and horizontal axes, respectively. Observe the remarkable similarity of the pictures with the reproduction of all subtle details in the structure of the SNAs.

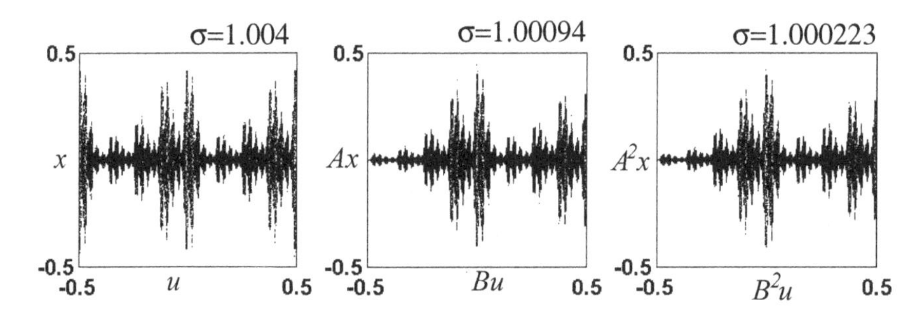

Fig. 7.9 Illustration of scaling of attractors in a vicinity of the critical point. Portraits of SNAs are shown at three parameter values; each next plot corresponds to a decrease of deflection from the critical point by factor $\delta = 4.2361$ and a scale change along the vertical and horizontal axes with factors $A = 1.3239$ and $B = -4.2361$, respectively. The negative factor b corresponds to inversion of the picture along the horizontal axis.

On the basis of the scaling argumentation, one can derive a relation for the extension of the size of SNA under increase of the parameter near the transition. For this, we have to use the scaling relations stated for a vicinity of $u = 0$. Apparently, these will be the largest absolute values of x in the attractor at phases near $u = 0$, as they have been growing on previous iterations without vanishing of the multiplicative driving term $\sin 2\pi u$ (the strong decay will occur just at the next step). So, under reduction of the distance from the critical point by $\delta = 4.236068$, the size of the attractor will decrease by factor $A = 1.32390$. That implies the power law relation $|x| \propto \Delta\sigma^{\gamma}$, $\gamma = \log A/\log\delta \approx 0.1943$ (the same index as obtained for the critical decay near zero phase).

In Fig. 7.10 (top panel) we present a 3D bifurcation diagram illustrating the birth of SNA, the absolute value of x versus phase u and parameter σ. One can clearly see that the growth of the size of the attractor has essentially different rates depending on the phase, but it is most rapid at $u = 0$. Fig. 7.10 (bottom panel) demonstrates the verification of the suggested relation: the maximal absolute value of x over the whole attractor is plotted versus parameter deflection from the critical point in double logarithmic scale. Observe that the numerical data follow well the predicted straight line with slope γ.

To conclude this section, we remark that the GOPY map suggested in the first original paper on SNA by Grebogi, Ott, Pelikan, and Yorke [1984]

$$x_{n+1} = 2\sigma \tanh x_n \sin 2\pi\theta_n \,, \quad \theta_{n+1} = \theta_n + \omega \pmod 1 \,, \tag{7.41}$$

demonstrates the same scaling regularities at the point of birth of SNA and for sure relates to the same universality class.

However, a more rich behavior is observed in a generalized model suggested by Glendinning [2004] (cf. (2.18))

$$x_{n+1} = 2\sigma(\cos 2\pi\theta_n + p) \tanh x_n \,, \qquad \theta_{n+1} = \theta_n + \omega \pmod 1 \,. \tag{7.42}$$

The parameter plane (σ, p) of this map (Fig. 7.11) contains three areas: the first one corresponds to the existence of the trivial attractor $x \equiv 0$, the second one to SNA, and the third one to a smooth torus attractor. The transition from the trivial attractor to SNA occurs at the bifurcation line $\sigma = 1$ in the interval of p from 0 to 1. The scaling regularities examined in this section take place at $p = 0$. On the remaining part of the bifurcation line $\sigma = 1$ they are modified because of the distinct location of zeros of the functions $Q(u)$ (see (7.32)). At the edge of the bifurcation line $p = 1$, a critical point of higher codimension is obviously present. This example shows that further investigations are desirable to reveal all possibilities that can occur in the case of the blowout transition to SNA in all detail under the most general circumstances.

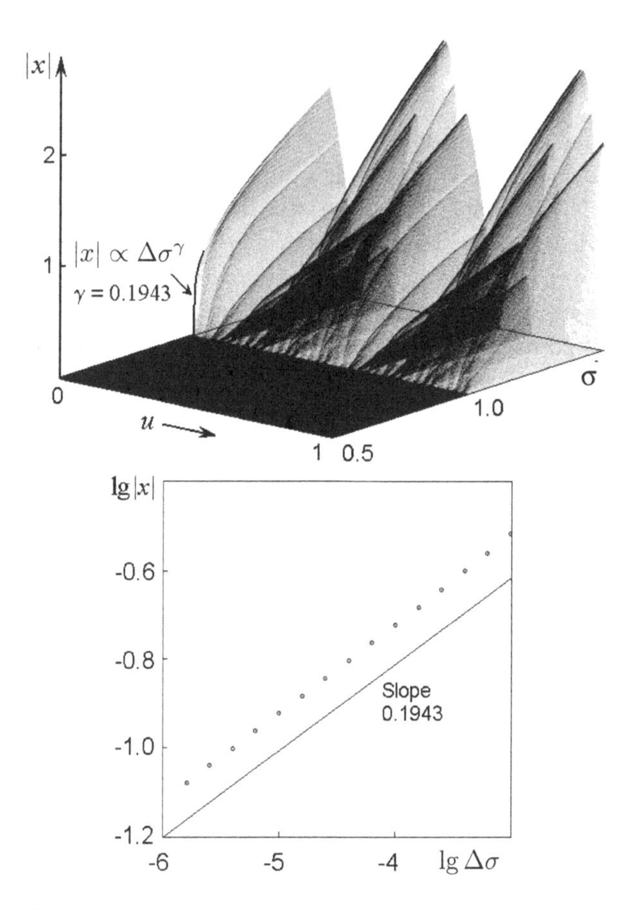

Fig. 7.10 A diagram of the absolute value of x versus phase u and parameter σ illustrating the birth of SNA in the model map (7.21) (top panel) and verification of the relation for the attractor size at $u = 0$ in double logarithmic scale (bottom panel).

7.6 RG analysis of the TDT critical point

In this section, we consider the torus-doubling terminal point (TDT) [Kuznetsov et al. 1998], which occurs in the forced quadratic map

$$x_{n+1} = a - x_n^2 + \varepsilon \cos 2\pi u_n\ , \qquad u_{n+1} = u_n + \omega \pmod 1\ . \qquad (7.43)$$

In contrast to the previous case of the blowout transition, a nontrivial numerical problem is to locate the critical point in the parameter plane accurately. To carry out this necessary preliminary step to the RG analysis, let us recall the definition of the TDT point.

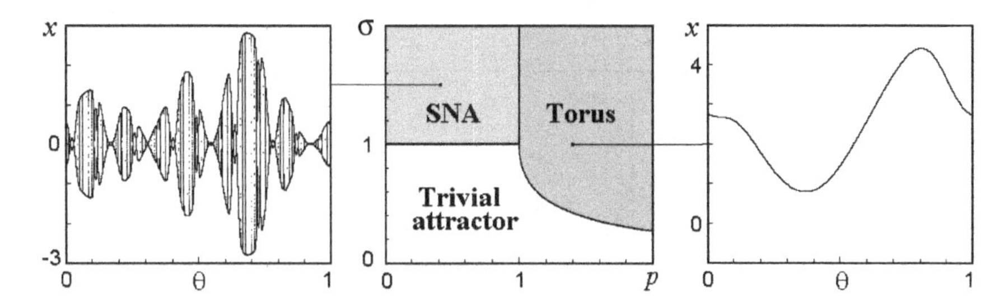

Fig. 7.11 Parameter plane of the modified GOPY map with quasiperiodic driving (7.42) with marked areas of the trivial attractor $x \equiv 0$, SNA, and torus. Examples of the phase portraits of SNA and torus are shown in the two side panels.

In the unforced map, at the first doubling bifurcation we have a fixed point at the stability threshold. For small amplitudes of driving, it turns into a smooth invariant curve placed entirely in the region $x > 0$ (Fig. 7.12). Let us suppose that we increase the amplitude and go along the bifurcation curve in the parameter plane, where the birth of the doubled torus T2 takes place from the original torus T1. In the course of this motion, the invariant curve becomes wider and wider, and the minimum value of x on it approaches, and finally reaches, zero. This event just corresponds to the TDT critical point. As we are at the bifurcation curve, a typical orbit on the torus is just at the instability threshold. On the other hand, the touch with zero implies the appearance of a superstable orbit on the torus, the one, which contains the parabola maximum $x = 0$.

Let us consider the situation in terms of rational approximations of the frequency parameter by ratios of Fibonacci numbers. For a rational frequency $\omega_k = F_{k-1}/F_k$, we get a cycle of period F_k instead of the T1 torus. Increasing the control parameter a we expect to see a bifurcation of this cycle at some parameter value that approximates the torus-doubling bifurcation. In fact, the bifurcation point for the rational frequency ω_k will depend on the initial phase u_0. Asymptotically, for $k \to \infty$, we can speak of a torus-doubling bifurcation only if the limit does not depend on u_0. Certainly, this is the case at small amplitudes ε. We may gradually increase ε and trace the torus-doubling curve as long as possible.

Let us state formal conditions of the situation in terms of rational approximations. At a fixed rational frequency $\omega_k = F_{k-1}/F_k$, we impose the following two requirements.

(i) For some initial phase $u_0 = u^0$ there exists a period-F_k cycle start-

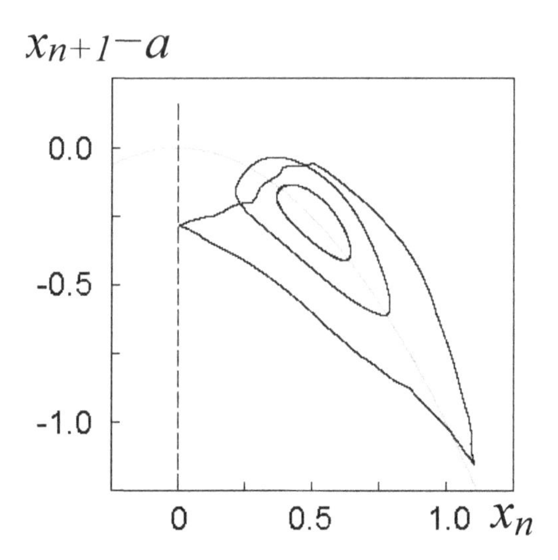

Fig. 7.12 Tori at the threshold of the doubling bifurcation: smooth tori for $\varepsilon = 0.1$, $a = 0.778791$ and $\varepsilon = 0.2$, $a = 0.824501$, and the critical torus for ε_c, a_c. For visual clarity, the vertical coordinate is shifted by a, so that the graph of the quadratic map has the same form for all parameter values. The critical torus touches the line $x = 0$.

ing from $x = 0$, and the derivative dx/du_0 vanishes. This cycle will be superstable, with zero Floquet multiplier. (The condition of zero derivative means a touch of the line $x = 0$ by the approximate torus: nearby orbits do not intersect this line.)

(ii) The minimum of the Floquet multiplier reached at some other initial phase u^1 equals (-1), i.e., for this phase the cycle of period-F_k is at the threshold of the period-doubling bifurcation. (Of course, this may be true only for the approximants with odd Fibonacci's denominators; therefore, we consider at this moment only them.) In the limit of the irrational frequency parameter, this corresponds to a condition of the torus-doubling bifurcation.

In Table 7.1 we summarize the computational data, which show an evident convergence. The estimate of the limit yields the location of the TDT point. Additionally, one can observe from the table a convergence of the phase sequence u^0 to a definite limit, which is also significant. (Note that the limiting value for u^1 is the same.) Our best result for these limits,

Table 7.1 Numerical data for the torus doubling terminal point in rational approximations for the driven quadratic map (7.43).

ω_k	a	ε	$2\pi u^0$
2/3	0.89313590	0.39045526	2.38814031
3/5	1.07633288	0.30511453	2.27096915
8/13	1.14077398	0.35637173	2.47438570
13/21	1.13453832	0.35326266	2.51704366
34/55	1.15587157	0.36021207	2.48341089
55/89	1.15364997	0.35864970	2.47129447
144/233	1.15790681	0.36022344	2.48295110
233/377	1.15720706	0.35995120	2.48531755
610/987	1.15807555	0.36024655	2.48328049
987/1597	1.15796494	0.36019031	2.48259605
2584/4181	1.15809462	0.36024806	2.48321966
4181/6765	1.15806354	0.36023723	2.48335560
10946/17711	1.15809658	0.36024835	2.48323605

improved with a technique based on further results of the RG analysis, is

$$a_c = 1.158096856726, \quad \varepsilon_c = 0.360248020507, \quad u_c = 0.3952188264. \tag{7.44}$$

At a point from the Table 7.1 with a rational frequency ω_k, we have simultaneously a superstable cycle at one phase, and a cycle at the period-doubling bifurcation threshold at another. Obviously, by an infinitesimal parameter shift from this point we can reach a situation that the cycle remains stable at one phase and becomes unstable at some other. This means inevitable occurrence of bifurcations depending on the phase of the external force, which persist at all levels of the rational approximation. According to the criteria formulated in Chapter 3 (see also [Pikovsky and Feudel 1995]), we conclude that a small parameter variation from the TDT point can give rise to an SNA. Additionally, numerical computations show that chaotic states appear also in a neighborhood of the TDT point. Finally, as it is clear from our way of reasoning, quasiperiodic regimes of type T1 and T2 also occur in its vicinity. Therefore, the significance of the TDT point is that a complete set of relevant dynamical regimes of the system occur arbitrarily close to it. It is natural to think that details and regularities of the dynamical behavior in this special local region of the parameter space are of principal importance for understanding general properties of the dynamics in the model under study, and, moreover, in the entire universality class associated with this type of critical behavior.

As the next step, let us consider a procedure aimed to reveal the nature of the solution of the RG equation responsible for the behavior at the

TDT critical point. First of all, we introduce a coordinate system on the phase plane (u, x) appropriate for constructing the RG transformation. It is natural to place the origin at $x = 0$, $u = u_c$. Indeed, the value $x = 0$ is an extremum of the forced one-dimensional map, obviously, a point of special significance, and u_c is the limit point of the phases for a sequence of cycles, superstable or bifurcating, on the approximated tori. We do not change the directions of the coordinate axes, so, the new coordinates are simply $X = x$ and $Y = u - u_c$. Now, we can compute terms of the functional sequence $g_k(X, Y)$ by virtue of straightforward iterations of the original maps (7.43) at the critical point. Given a sufficiently large Fibonacci number F_k (say, 233), we first set the initial condition $x_0 = 0$, $u_0 = u_c$ and iterate the mapping F_k times. The resulting $x_{F_k} = x^0_{F_k}$ determines a normalization factor used later. Now, to obtain the value of $g_k(X, Y)$ for some particular X and Y, we iterate the map (7.43) again, but with initial conditions $x_0 = x^0_{F_k} X$, $u_0 = (-\omega)^k Y$. After F_k iterations we obtain x_{F_k} which after a normalization yields $g_k(X, Y) = x_{F_k}/x^0_{F_k}$. Such computations show that for high-order Fibonacci numbers the functions $g_k(X, Y)$ form a sequence of period 3: $g_k(X, Y) \approx g_{k+3}(X, Y)$. Therefore, we conclude that this is the period-3 solution of the RG equation, which is responsible for the critical behavior at the TDT point. Moreover, the computations yield an estimate for the factor of renormalization for the x variable, $\alpha^3 \cong x_{F_k}/x_{F_{k+3}} \cong 3.96$.

The next step is to get an accurate numerical solution of the functional equation (7.6). For this, we approximate the functions of two variables constituting the period-3 cycle $g_1(X, Y) \to g_2(X, Y) \to g_3(X, Y) \to g_1(X, Y)$ by finite polynomials with unknown coefficients, and one more unknown quantity is the factor α. The polynomials contain odd and even powers of Y and even powers of X. One of the three functions is assumed to be normalized to unity, say $g_1(0, 0) = 1$. Then, the procedure of RG transformation is organized as a computer program operating with the coefficients of polynomials. It corresponds to an implicitly defined finite set of nonlinear algebraic equations. This set is solved by means of a multi-dimensional quasi-Newton method. The crucial point is to have a good initial guess for the unknowns; we take them from the numerical estimates for the functions g_m and the constant α, described in the previous paragraph.

The final result is a representation of the functional pair $\{g_1(X, Y), g_2(X, Y)\}$ in a form of polynomials in X and Y. (The third function can be computed easily from the functional composition rule following from the basic equation (7.6).) The concrete expressions are too tedious to present them here (see the table of coefficients published in [Bezruchko et al. 1997] and on the web page http://www.sgtnd.narod.ru/science/alphabet/eng/goldmean/tdt.htm). Figure 7.13 shows a series of 3D plots illustrating this solution. The most accurate estimate for the rescaling factor obtained from the computations is

$$\alpha = 1.58259341\ldots. \tag{7.45}$$

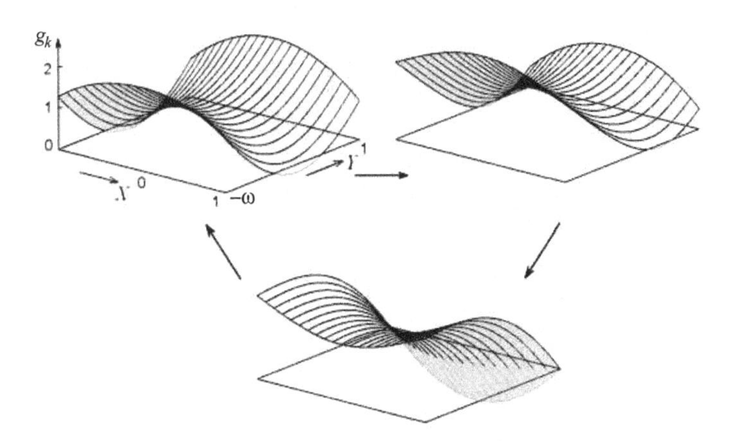

Fig. 7.13 3D plots of functions $g_k(X,Y)$ constituting the period-3 cycle of the RG equation associated with the TDT critical point

Function $g_k(X,Y)$ determines the rescaled evolution operators for F_{k+3m} iterations of the driven quadratic map at the TDT critical point (in fact, in the asymptotics $m \to \infty$). A cyclic repetition of these functions implies the presence of scaling regularities, i.e. similarity of the dynamics on different time scales. Three steps of the RG procedure correspond to an increase of the time scale by a factor $\tau_k = F_{k+3}/F_k$, asymptotically $\tau = 1/\omega^3 = 4.236068$, and to a rescaling of x and u variables by factors $A = \alpha^3 = 3.96376647$ and $B = 1/(-\omega)^3 = -4.236068$, respectively. This means that iterations of the forced quadratic map at the critical values of a and ε starting from some x and u, and those starting from x/A and $u_c + (u - u_c)/B$ demonstrate similar dynamics. In the second case the time scale is greater by τ, and the phase space scales for x and u are less, by factors A and B, respectively.

While we move on the parameter plane from a small driving amplitude to the TDT point, following the torus-doubling bifurcation line, the invariant curve remains smooth (Fig. 7.12). As we reach the TDT point, the invariant curve becomes a fractal-like. We call it the *critical attractor* and show it in Fig. 7.14 in coordinates (u, x).

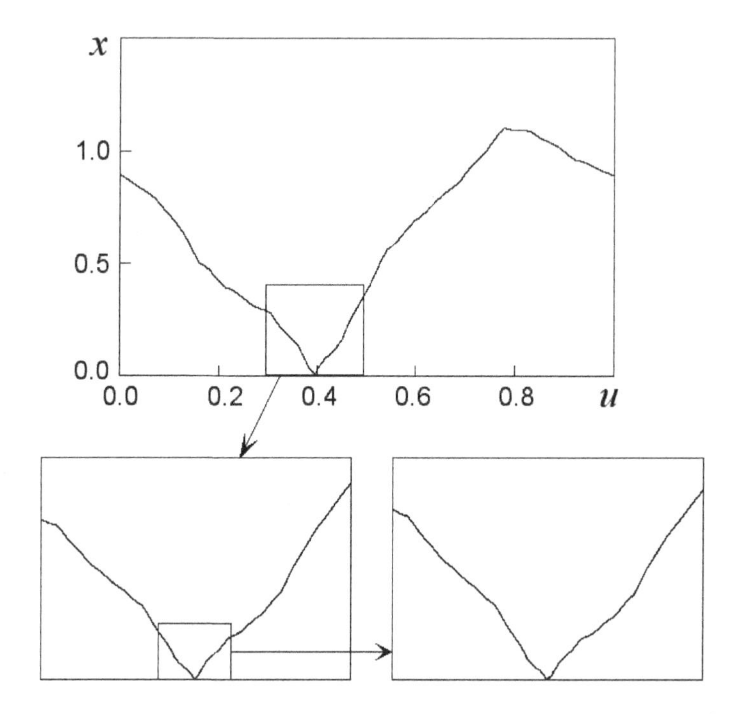

Fig. 7.14 Attractor of the forced quadratic map at the TDT critical point (the top panel) and illustration of the basic local scaling property: the structure reproduces itself under magnification with factors $A = \alpha^3 = 3.96376$ and $B = \beta^3 = -4.2360$ along the vertical and the horizontal axes, respectively. The negative scaling factor B implies reflection along the horizontal axis

As it follows from the RG analysis, the critical attractor must demonstrate self-similarity, and we can explicitly formulate this property in a vicinity of the point $x = 0$, $u = u_c$ ('*the scaling center*'). If we rescale $X = x$ and $Y = u - u_c$, respectively, by A and B, the curve representing the attractor must be locally invariant under this transformation. As can be seen from Fig. 7.14, this is indeed the case: the picture inside a selected box reproduces itself under subsequent magnifications by the factors A and B. To possess the stated scaling property, the invariant curve must behave locally as $x \propto |\Delta u|^{\kappa}$ with $\kappa = \log \alpha / \log |\beta| \cong 0.954$. The exponent κ is close to 1, so visually the curve looks like a line broken at the point of singularity. It is worth noting that due to ergodicity of the quasiperiodic motion, the singularity at the origin implies the presence of singularities of the same type at all points, to which it is mapped under subsequent iterations of the map. This is a dense set of points on the invariant curve. As κ is less than 1, the invariant curve in nowhere differentiable. Thus, the conclusion on a fractal nature of the critical attractor follows directly from the results of the RG analysis.

The stability analysis in terms of the linearized RG equation for the period-3 cycle solution responsible for the TDT criticality gives rise to the following eigenvalue problem:

$$\begin{aligned} h_3(X,Y) = &\ \alpha g_1'(\alpha^{-1}g_2(X/\alpha,\ -Y\omega),\ \omega^2 Y+\omega)h_2(X/\alpha,\ -Y\omega) \\ &+\alpha^2 h_1(\alpha^{-1}g_2(X/\alpha,\ -Y\omega),\ \omega^2 Y+\omega), \\ \delta h_2(X,Y) = &\ \delta\alpha g_3'(\alpha^{-1}g_1(X/\alpha,\ -Y\omega),\ \omega^2 Y+\omega)h_1(X/\alpha,\ -Y\omega) \\ &+\alpha^2 h_3(\alpha^{-1}g_1(x/\alpha,\ -y\omega),\ \omega^2 Y+\omega), \\ \delta h_1(X,Y) = &\ \alpha g_2'(\alpha^{-1}g_3(X/\alpha,\ -Y\omega),\ \omega^2 Y+\omega)h_3(X/\alpha,\ -Y\omega) \\ &+\alpha^2 h_2(\alpha^{-1}g_3(X/\alpha,\ -Y\omega),\ \omega^2 Y+\omega), \end{aligned} \tag{7.46}$$

where δ is the eigenvalue, and $h_{1,2,3}(X,Y)$ are functions constituting the eigenvector. The numerical solution of this problem is based on the approximation of $h_{1,2,3}(X,Y)$ by finite polynomials, with substitution of $g_{1,2,3}(X,Y)$ and α, which are already known. This way, we reduce the functional eigenproblem to that for a finite matrix, which allows a solution with standard methods of linear algebra. After excluding irrelevant eigenvalues (which are either less than 1 in modulus, or associated with infinitesimal variable changes or with departures from the commutative subspace) we have the following two eigenvalues in the rest:

$$\delta_1 = 10.5029\ldots \qquad \text{and} \qquad \delta_2 = 5.1881\ldots \tag{7.47}$$

These are the scaling factors responsible for properties of self-similarity in the arrangement of the parameter plane near the TDT critical point.

To observe the scaling regularities in a vicinity of the critical point we need to define an appropriate local coordinate system ('scaling coordinates'). As the origin we naturally take the TDT point itself. Then, the coordinate axes should be directed in such a way that a shift along the first one generates the mode of perturbation of the evolution operators associated with δ_1, and the mode with δ_2 is produced with a shift along the second axis. In such a coordinate system the configuration of regions in the parameter plane near the TDT point will be self-similar and will reproduce itself locally under scale change along the coordinate axes by δ_1 and δ_2, respectively.

In fact, an axis, corresponding to the larger eigenvalue δ_1 may be defined almost arbitrarily, say, as a direction along the a axis, but the second one must be selected very carefully to exclude a contribution of the first eigenmode into the solution on this axis. As is found numerically, for our model (7.43) the appropriate new coordinates (c_1, c_2) and the original parameters (a, ε) are related as follows:

$$a = a_c + c_1 + c_2 \,, \quad \varepsilon = \varepsilon_c + 0.3347 c_2 \,. \tag{7.48}$$

(For our purposes of demonstrating scaling in the case of the TDT point it is sufficient to use a linear parameter change because of the relation between the relevant eigenvalues: $\delta_2 < \delta_1$ and $\delta_2 > \delta_1^m$ for $m = 2, 3, \ldots$). Figure 7.15 shows a chart of dynamical regimes for the driven quadratic map (7.43) near the TDT point in scaling coordinates for several steps of subsequent magnification.

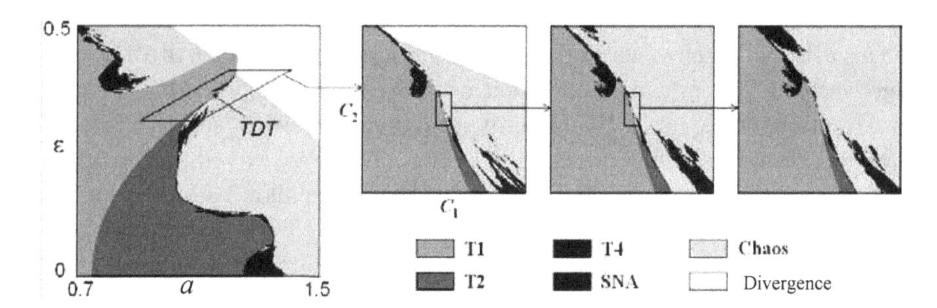

Fig. 7.15 Chart of the dynamical regimes on the parameter plane of the quasiperiodically driven quadratic map and a sequence of fragments for several steps of magnification of a vicinity of the TDT critical point in scaling coordinates, with factors δ_1 and δ_2 along the horizontal and vertical axes, respectively. The dark gray area corresponds to localized attractors with negative Lyapunov exponent, light gray relates to chaos and white to divergence.

Figure 7.16 illustrates the scaling property with Lyapunov charts, where the values of the nontrivial Lyapunov exponent are shown by gray scales. The Lyapunov exponent is inversely proportional to the characteristic time scale. Therefore, to outline the similarity of the pictures, the coding rule is redefined at each next level of magnification in such way that it corresponds to rescaling of the Lyapunov exponent by factor τ.

One can observe a good correspondence of the pictures at higher levels of magnification under the scale change by factors δ_1 and δ_2 along the axes of the scaling coordinates. It means that the configuration of the regions in

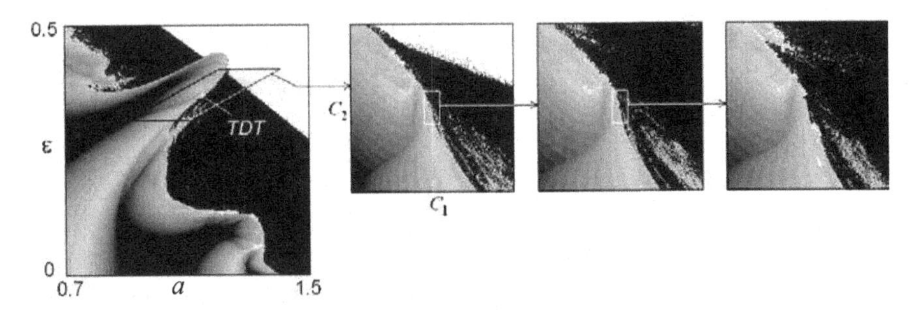

Fig. 7.16 Scaling in the vicinity of the critical point illustrated by the Lyapunov charts. The Lyapunov exponent is coded in gray scale. Left panel: a region around the TDT point in natural coordinates. The interior of the parallelogram is shown separately in scaling coordinates for several steps of magnification with factors δ_1 and δ_2 along the horizontal and vertical axes, respectively. The gray coding for the Lyapunov exponent is scaled with the factor τ.

the parameter plane indeed obeys the expected property of self-similarity. In particular, it follows that all the regimes: T1, T2, SNA, chaos, occur arbitrarily close to the critical point.

As expected, the regularities intrinsic to the dynamics at the TDT critical point and in its vicinity are universal; in particular, this relates to the estimated numerical values of the scaling constants. Therefore, this type of critical behavior has to occur in many nonlinear dissipative systems manifesting the period-doubling bifurcation cascade, in the presence of an additional periodic forces with incommensurate frequencies (the golden-mean frequency ratio). Heuristically, the hypothesis of universality follows from the RG argumentation, in the same spirit as the Feigenbaum universality for the classic period doubling [Feigenbaum 1979, 1983].

The essential feature of the TDT universality class is a particular parameter plane topography in a vicinity of the critical point. Depicted in scaling coordinates it will be universal, and for this reason it deserves special attention. The consideration reveals some links with bifurcation scenarios in quasiperiodically forced systems mentioned in Chapter 6.

In the left top panel of Fig. 7.17 we reproduce in scaling coordinates a sketch of the parameter plane arrangement near the TDT point. On the roads marked (a), (b), and (c) one can observe the bifurcation scenarios involving SNA; see other panels of the figure. In the bifurcation diagrams, the horizontal axis corresponds to the parameter variation along the respective path in the parameter plane, and the vertical axis to the dynamical variable x. To produce distinguishable diagrams, we use a gray coding rule:

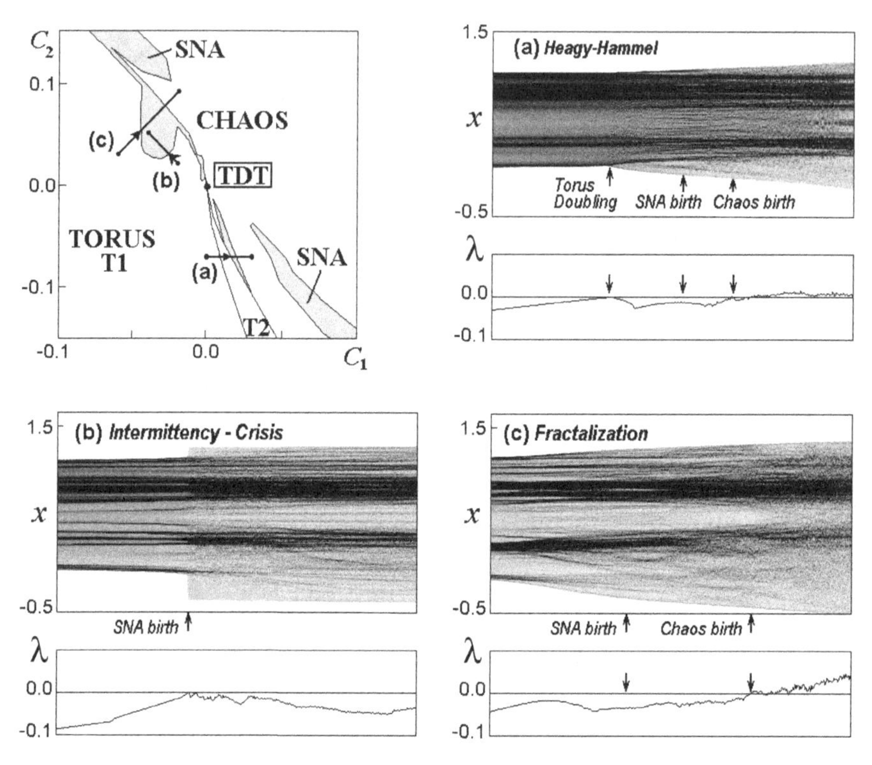

Fig. 7.17 The left top panel is a sketch of the parameter plane topography near the TDT point in scaling coordinates; the areas of torus T1, doubled torus T2, SNA, and chaos are shown. Panels (a), (b), (c) present the bifurcation diagrams (x variable versus parameter) and Lyapunov exponent plots along the paths in the parameter plane marked with the respective letters. The path (a) corresponds to the Heagy-Hammel scenario of birth of SNA (via collision of the wrinkled doubled torus $T2$ with the parent torus T1), along the path (b) the SNA appears via the intermittency transition examined by Prasad et al. [1997] and by Kim et al. [2003b], and on the path (c) via gradual fractalization as suggested by Nishikawa and Kaneko [1996].

each time a point hits a definite pixel, the gray code number is increased by 1. In addition, we present respective plots for the Lyapunov exponent; recall that the negative values are associated with a torus-attractor, or an SNA, and the positive ones with chaos.

The road marked (a) corresponds to the Heagy-Hammel scenario [Heagy and Hammel 1994]. On this path, the parent torus T1 becomes unstable and gives birth to the doubled torus T2 via a torus-doubling bifurcation. Then, the torus T2 grows in width, becomes wrinkled, and touches the unstable parent torus T1. This is the moment of birth of the SNA. A further motion along this path gives rise to chaos (observe the appearance of the positive Lyapunov exponent). The road (b) corresponds to the birth of SNA via intermittency in a transition examined e.g. by Prasad et al. [1997] and Kim et al. [2003b]. It looks like a sharp increase of the attractor size, but beyond the transition the system spends the dominating part of time in a vicinity of the former attractor. The mechanism of this transition resembles an interior crisis — the collision of an attractor with a saddle invariant set ("ring-shape sets" in the rational approximation approach, in the terminology of [Kim et al. 2003b]). Finally, the road (c) corresponds to the scenario of gradual fractalization of a torus-attractor suggested by Nishikawa and Kaneko [1996].

We emphasize that the sketched picture appears repeatedly in smaller scales under magnification by factors δ_1 and δ_2 along the axes of scaling coordinates, with all mentioned elements of the universal pattern of the parameter plane topography. Thus, we conclude again that the TDT point is like an "organizing center" of the whole pattern: any small vicinity of it contains all the main dynamical regimes relevant for the system – tori T1 and T2, SNA, chaos.

7.7 RG analysis of the TCT critical point

In this section, we turn to the RG analysis for the torus collision terminal point (TCT) [Kuznetsov et al. 2000]. The critical points of this type occur in the driven quadratic map, the driven supercritical circle map, and in similar models. To perform the RG analysis, it is more convenient to use the first model, which is simpler. However, for illustrations and classification of dynamical behaviors, the circle map model is more appropriate, because no divergence of iterations ever happens, in contrast to the quadratic map. In many respects, the scheme of reasoning will be very close to that developed in concern with the TDT point in the previous section.

Let us start with the procedure of accurate location of the TCT point in the parameter plane of the driven quadratic map

$$x_{n+1} = a - x_n^2 + \varepsilon \cos 2\pi u_n\,, \qquad u_{n+1} = u_n + \omega \pmod 1\,. \tag{7.49}$$

In the unforced map, a tangent bifurcation, that is a collision of a stable fixed point with an unstable one, takes place at $a = -0.25$ and at $x = -0.5$. This bifurcation corresponds to the lower border of the stability interval of a for the fixed point. For small amplitudes of driving, a fixed point transforms into a smooth closed invariant curve, and the bifurcation consists in a collision of two such curves, one born from a stable, and another from an unstable fixed point, with their subsequent disappearance. At the bifurcation they coincide, and there is a single smooth invariant curve placed entirely in the region $x < 0$ (Fig. 7.18). If we increase the amplitude of forcing and go along the bifurcation curve in the parameter plane, the interval occupied by the invariant curve becomes wider, and the maximum value of x on it approaches zero. Finally, it reaches zero. This event corresponds to the terminal point of the bifurcation curve, the TCT critical point. As we stay on the bifurcation curve, a typical orbit on the torus is just at the instability threshold. However, a touch with zero implies that a superstable orbit, which contains the parabola maximum x=0, appears among other trajectories on the torus.

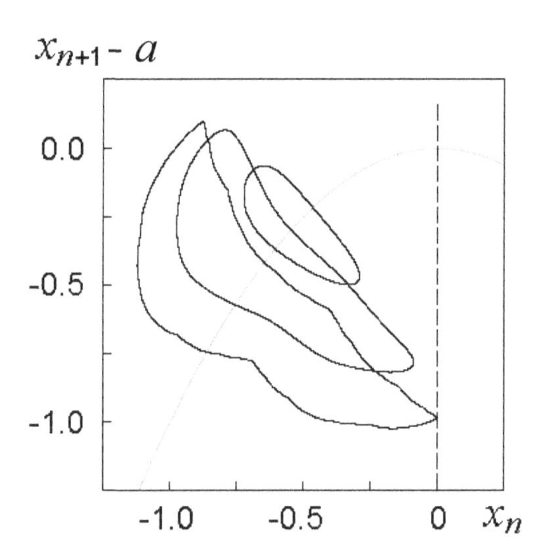

Fig. 7.18 The invariant curves at the torus collision bifurcation: smooth tori for $\varepsilon = 0.4$, $a = -0.22696$ and $\varepsilon = 0.8$, $a = -0.15716$, and the critical torus at ε_c, a_c. The plot is drawn with a shift of the vertical coordinate by a, so that the graph of the quadratic map has the same form for all parameter values. The critical torus touches the line $x = 0$.

For a rational frequency $\omega_k = F_{k-1}/F_k$, instead of the invariant curve, we get a cycle of period F_k. It depends on the initial phase u_0, which may be regarded as an additional parameter. Let us require that the set of orbits with different initial phases, which approximates the torus, touches $x = 0$. It means that at some phase $u_0 = u^0$ there exists a cycle of period-F_k starting from $x = 0$, and the derivative dx/du_0 vanishes. (Zero derivative means a touch of the line $x = 0$ by the approximate torus: nearby orbits do not intersect this line.) Simultaneously, we demand the collision bifurcation to occur, i.e. the maximum of the Floquet multiplier equals 1 for the period-F_k cycle at some other initial phase $u_0 = u^1$.

At each level of the rational approximation, we can estimate numerically a and ε, and the phases u^0 and u^1, at which the conditions are satisfied. In Table 7.2 the numerical data are summarized. One can observe an evident convergence with the increase of the order of approximation. The limits for a and ε determine the location of the TCT point in the parameter plane. Our best result obtained from computations with high accuracy (60-digit precision, at levels of Fibonacci numbers up to 514229...3524578) is

$$a_c = -0.09977122895\,, \quad \varepsilon_c = 1.01105609099\,, \text{ and } \quad u_c = 0.53372941325\,, \tag{7.50}$$

where u_c is a common limit for the phase sequences u^0 and u^1.

Table 7.2 Numerical data for the torus collision terminal point in rational approximations for the driven quadratic map (7.49).

ω_k	a	ε	u^0
8/13	-0.106863851767	0.989187201082	0.505473909903
13/21	-0.104845946249	0.993122021583	0.551881855417
21/34	-0.102132053793	1.003742152765	0.523043122989
34/55	-0.101301679580	1.005940852360	0.540532546847
55/89	-0.096519589548	1.008703220940	0.529671825167
89/144	-0.096227513110	1.009565266966	0.536295837724
144/233	-0.096003370260	1.010319151502	0.532185483229
233/377	-0.099907304210	1.010615844857	0.534709662240
377/610	-0.099842405051	1.010828952123	0.533141299658
610/987	-0.09981191332417	1.01096505020861	0.53409804003715
987/1597	-0.09979689883669	1.01098675212306	0.53350525381290
1597/2584	-0.09978341968243	1.01101690728124	0.53386947751449
2584/4181	-0.09977779656842	1.01103504807250	0.53364394312993
4181/6765	-0.09977488690490	1.01104434510749	0.53378268847958
6765/10946	-0.09977321340337	1.01104972857703	0.53369681588723

The TCT critical point occurs also in the forced supercritical circle map

$$\begin{aligned} x_{n+1} &= x_n + c - a\sin 2\pi x_n + \varepsilon\cos(2\pi u) \pmod 1\,, \\ u_{n+1} &= u_n + \omega \pmod 1\,. \end{aligned} \tag{7.51}$$

'Supercritical' means that the parameter a is larger than $1/2\pi$, and the mapping for x in non-invertible. For definiteness, we fix $a = 2.5/2\pi$. Locally, near the extremum at $x^0 = \arctan(\sqrt{a^2 - 1/(4\pi^2)})$, the function on the right hand side in the equation looks similar to the parabola map, so it is natural that the same type of criticality happens here. The condition is that the invariant curve at the threshold of the collision bifurcation touches x^0. In Table 7.3 we present numerical data for the circle map at rational approximations of the frequency parameter (analogous to those in Table 7.2 for the quadratic map). Observe the convergence to definite limits determining location of the TCT point. The best estimates obtained from computations with 20-digit precision up to Fibonacci numbers 46368...317811 yield

$$c_c = 0.377866239\,, \qquad \varepsilon_c = 0.132566321\,, \qquad u_c = 0.284109286\,. \quad (7.52)$$

Table 7.3 Numerical data for the torus collision terminal point in rational approximations for the driven circle map (7.51) at $a = 2.5/(2\pi)$.

ω_k	c	ε	u^0
8/13	0.37881706168188	0.12963650656581	0.25582995050593
13/21	0.37853943959631	0.13019323399852	0.30220519948417
21/34	0.37818176558166	0.13158976184330	0.27342021471748
34/55	0.37806966405365	0.13188906990693	0.29089396540962
55/89	0.37796607502727	0.13225282828857	0.28005156151726
89/144	0.37796680039671	0.13236878805042	0.28666996177228
144/233	0.37789716819380	0.13246827351501	0.28256543113078
233/377	0.37788430788852	0.13250795751162	0.28507911784015
377/610	0.37787571332547	0.13253613129985	0.28352121995658
610/987	0.37787164279323	0.13254894376094	0.28447737339461
987/1597	0.37786912162019	0.13255711167252	0.28388514644255
1597/2584	0.37786785844262	0.13256112439438	0.28424919682601

Now, to reveal the nature of the solution of the RG equation (7.6) associated with the TCT criticality, one can try to compute the terms of the functional sequence $g_k(X,Y)$ via iterations of a model map at the critical point. At this place, it is more convenient to deal with the quadratic map. Given a sufficiently large Fibonacci number F_k, we first set the initial condition $x_0 = 0$, $u_0 = u_c$ and from iterations of the map get $x_{F_k} = x^0_{F_k}$. Then, to obtain $g_k(X,Y)$ for some particular X and Y, we again iterate the map (7.49), but with initial conditions $x_0 = x^0_{F_k}X, u_0 = u_c + (-\omega)^k Y$. After F_k iterations we obtain the resulting x_{F_k} and set $g_k(X,Y) = x_{F_k}/x^0_{F_k}$. Such computations show that the sequence of functions tend to a definite limit, $g_k(X,Y) \to g(X,Y)$. This is a fixed point of the RG equation (7.6), i.e.

$$g(X,Y) = \alpha^2 g(\alpha^{-1} g(X/\alpha,\ -\omega Y),\ \omega^2 Y + \omega)\ . \tag{7.53}$$

Here the scaling factor α, as estimated from the computations, is $\alpha \cong x^0_{F_k}/x^0_{F_{k+1}} \cong 1.71$. (In fact, the convergence is rather slow, and to get a really good approximation it is necessary to use either sufficiently large Fibonacci numbers and/or to reveal and account for the character of the convergence; see [Kuznetsov et al. 2000] for some details.)

The next step is to obtain an accurate numerical solution of the functional equation (7.6). One possible approach is to approximate the function $g(X,Y)$ by a finite polynomial containing odd and even powers of Y and even powers of X, and to search numerically for a set of coefficients of this polynomial satisfying Eq. (7.6) with the best possible accuracy. A straightforward realization of this idea appears not to be feasible, and some tricks are necessary. First, we select a restricted domain of the definition for the function g in the (X,Y) plane. The condition is that for any point (X,Y) of this domain D the points $(X/\alpha, -\omega Y)$, and $(\alpha^{-1}g(X/\alpha, -\omega Y), \omega^2 Y + \omega)$ (see the right hand side of Eq. (7.53)) belong to D. From the approximate data for $g(X,Y)$, one can check that the domain

$$D : \{-0.1 + 0.9Y < |X| < 0.1 + 0.9Y, -\omega < Y < 1\} \tag{7.54}$$

is appropriate. For a representation of the function in D an expansion in orthogonal Chebyshev polynomials in two arguments has been applied, and by a Newton method the coefficients of this expansion have been computed. As the initial guess, a function obtained numerically from iterations of the quadratic map at the TCT point was used. This procedure results in a solution of Eq. (7.6) with precision of order 10^{-7}. To compute the function outside the region D, one has to use the functional equation (7.6) and to construct appropriate compositions of the numerically obtained function under the requirement that the argument values, at which the computations are performed, relate to D. A plot of the fixed-point function is shown in Fig. 7.19. The accurate estimate for the scaling factor obtained from the solution of the functional equation is

$$\alpha = 1.7109605\ . \tag{7.55}$$

Up to a characteristic scale, the function $g(X,Y)$ determines asymptotically the evolution operators for F_m iterations of the original driven

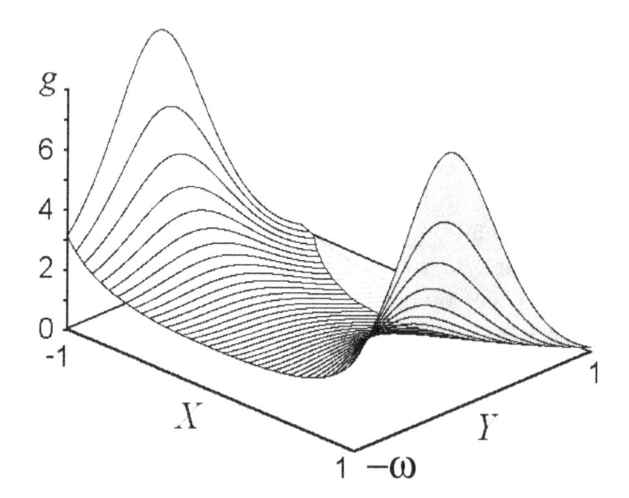

Fig. 7.19 3D plot of the function $g(X,Y)$ representing the fixed point of the RG equation associated with the TCT critical point.

quadratic map at the TCT critical point. It implies the presence of similarity of the dynamics on the respective time scales as well as scaling regularities.

One step of the RG procedure corresponds to an increase of the time scale by factor $\tau_k = F_{k+1}/F_k$, asymptotically $\tau = 1/\omega = 1.618034$, and to a decrease of the scales for variables $X = x$ and $Y = u - u_c$ by factors $\alpha = 1.7109605$ and $\beta = 1/(-\omega) = -1.618034$, respectively.

From Fig. 7.18 one can observe that the attractor at the TCT critical point looks as a fractal-like curve. To reveal its scaling property, let us consider a plot of the critical attractor in coordinates (u, x). If we rescale $X = x$ and $Y = u - u_c$ by factors α and β, respectively, then the dynamics is expected to be similar, but with an increase of the characteristic time scale by τ. Hence, the curve representing the critical attractor must be invariant under this transformation. Figure 7.20 demonstrates that this is indeed the case: The picture inside the selected box reproduces itself under subsequent magnifications (with inversion with respect to the horizontal axis, due to the negative scaling factor). This scaling property near the origin implies that the critical curve must behave locally as $x \propto |\Delta u|^{\kappa}$ with $\kappa = \log\alpha/\log|\beta| \cong 1.117$. The power is close to one, so visually the curve looks like a line, broken at the point of singularity. Singularities of the same type will be present also at all points, to which the origin is mapped under subsequent iterations. Due to ergodicity of the quasiperiodic motion, this is a dense set of points over the invariant curve, and this explains the fractal-like shape. In fact, the singularity is weak: κ is slightly larger than 1. Thus, the invariant curve is differentiable, although not twice differentiable.

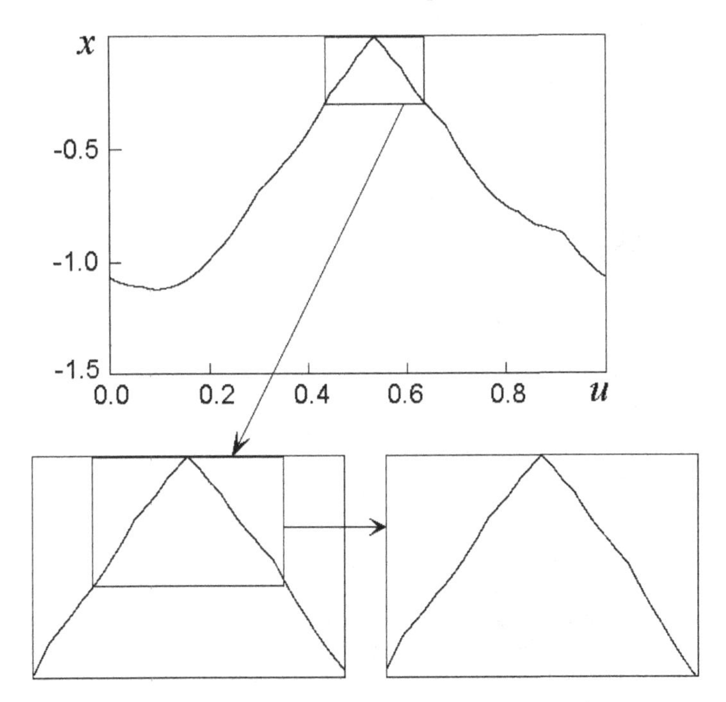

Fig. 7.20 Scaling of the critical torus for the driven quadratic map in the phase plane (u, x) at the TCT point. The top panel shows the whole picture, the next one presents the enlargement of the selected box, and the last – enlargement of the box from the previous one, under magnification by $\alpha = 1.7109$ for the vertical axis, and $\beta = (-\omega)^{-1} = -1.6180$ for the horizontal axis. Negative value of β means that the magnification is accompanied by reflection along the horizontal direction.

Fig. 7.21 presents an analogous verification for the scaling property of the critical attractor at the TCT point of the driven circle map. Here the main singularity is placed at $x^0 = \arctan(\sqrt{a^2 - 1/(4\pi^2)})$. This is a minimum of the right-hand side (in contrast to the considered logistic map, which has a maximum); for this reason the picture is upside down.

Before discussing the scaling regularities intrinsic to a vicinity of the TCT point, we have to remind that this is where different bifurcation lines meet, one corresponds to the smooth tori collision, and another to the intermittent transition (see Section 7.3). Beyond crossing one of these bifurcation lines, the localized attractor disappears, but in the part of the

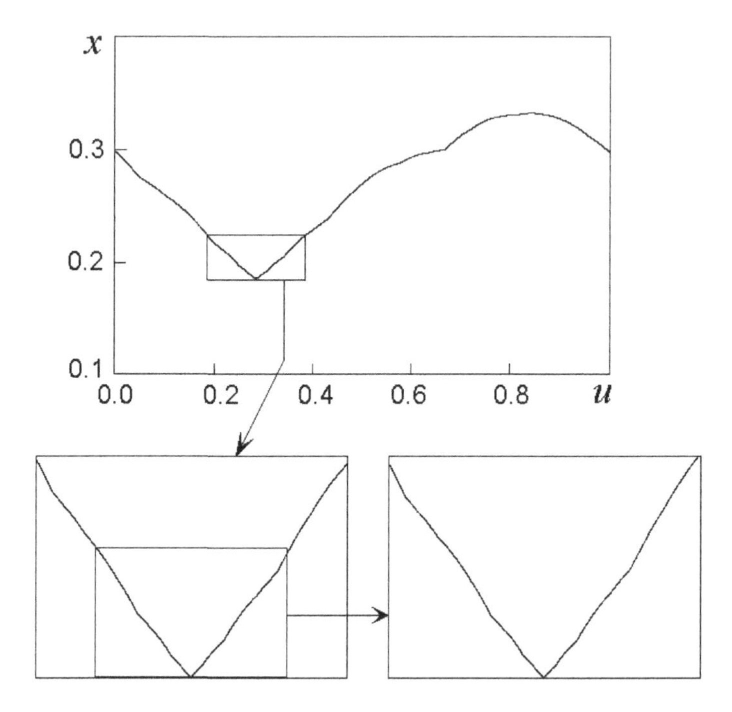

Fig. 7.21 Scaling of the critical torus for the driven circle map in the phase plane (u, x) at the TCT point. The top panel shows the whole picture, the next one presents the enlargement of the selected box, and the last – enlargement of the box from the previous one, under magnification by $\alpha = 1.7109$ for the vertical axis, and $\beta = (-\omega)^{-1} = -1.6180$ for the horizontal axis.

phase space of its former existence the trajectories travel for a very long time: the closer to the bifurcation, the larger the characteristic time. From the global point of view, there are two possibilities. First, it may occur that after the passage through the mentioned part of the phase space the orbit never returns back (this is the case for the quadratic map, where the divergence takes place). Alternatively, it may happen that a re-injection mechanism exists, due to which the orbit returns into that part of the phase space again and again (this is the case for the circle map). Then, beyond crossing the bifurcation borders of the area of torus stability one will observe a transition to SNA or to chaos via intermittency. These dynamical regimes, however, are not bounded in the domain of the former existence of the localized attractor, and, hence, the conclusions from the analysis in terms of the linearized RG equation are not applicable for them, at least not in a straightforward manner. (Here an analogy may be useful with the analysis of the classic intermittency of Pomeau and Manneville [1980b], where the RG approach relates only to laminar phases, while the properties for the turbulent bursts are postulated, as an additional component of the theory.) Having in mind these restrictions, we assume further that the second alternative takes place, and turn now to analyses of the perturbations for the fixed-point solution of the RG equation.

Let us substitute $g_k(X,Y) = g(X,Y) + \delta^k h(X,Y)$ into Eq. (7.6), and in the first order in h we get the eigenvalue problem for the linearized functional equation

$$\begin{aligned}\delta^2 h(X,Y) = \alpha\delta g'(\alpha^{-1}g(X/\alpha,\,-\omega Y),\,\omega^2 Y+\omega)h(x/\alpha,\,-\omega Y)\\ +\alpha^2 h(\alpha^{-1}g(X/\alpha,\,-\omega Y),\,\omega^2 Y+\omega)\,.\end{aligned} \tag{7.56}$$

As we locate the TCT point tuning two free parameters, we expect to have two relevant eigenvalues, namely, δ_1 and δ_2, which are larger than 1 in modulus. Then the asymptotic behavior of the infinitesimal perturbation to the fixed-point solution will contain the corresponding two eigenvectors and for the perturbed evolution operator we have to write

$$g_m(X,Y) \cong g(X,Y) + C_1(\varepsilon,b)\delta_1^m h_1(X,Y) + C_2(\varepsilon,b)\delta_2^m h_2(X,Y)\,. \tag{7.57}$$

Here C_1 and C_2 are coefficients, which depend on the parameters of the original map and vanish at the critical point. The numerical solution of the linearized equation (7.56) is based on the approximation of the function $h(X,Y)$ by finite Chebyshev polynomial expansions, with substitution of $g(X,Y)$ and α, which are already computed. This way, we reduce the functional equation to an eigenproblem for a finite matrix, which allows a solution by standard methods of linear algebra. After excluding the eigenvalues, which are less than 1 in modulus, those associated with infinitesimal variable changes, and with departures from the commutative subspace, only two relevant eigenvalues remain in the rest:

$$\delta_1 = 3.600810...\,,\quad \delta_2 = 1.828329...\,. \tag{7.58}$$

To state the basic scaling property, let us suppose that we consider a dynamical regime at a point (ε, c), which corresponds to some values of the coefficients $C_1 = C_1^0$ and $C_2 = C_2^0$ in (7.57). If we now take such a point (ε', c') that the coefficients will be equal to $C_1 = C_1^0/\delta_1$ and $C_2 = C_2^0/\delta_2$, then the renormalized evolution operator for F_{k+1} iterations at the new point will be the same as the evolution operator for F_k iterations at the old point (see (7.57)).

For convenience, we could introduce a special local coordinate system (*scaling coordinates*) near the critical point; for this we simply regard C_1, C_2 as the coordinates. Then, a simultaneous scale change along the coordinate axes by factors δ_1 and δ_2, respectively, corresponds to the condition of similarity of the evolution operators. Unfortunately, we do not know explicit expressions for C_1, C_2 in terms of the parameters of the original map, so the problem has to be resolved numerically, with a sufficient accuracy.

We follow an alternative approach, constructing a coordinate system (c_1, c_2) with the origin at the critical point. One coordinate axis, corresponding to the larger scaling factor δ_1, may be directed almost arbitrarily. A shift along this direction has to contribute to the coefficient C_1 in (7.57); so, the only condition is that it must be transversal to a curve on the parameter plane defined by an equation $C_1(\varepsilon, c) = 0$. By contrast, the second coordinate must be defined carefully, because its contribution to the senior eigenvector should be excluded. For this, the curve $C_1(\varepsilon, c) = 0$ in new coordinates has to be a coordinate curve, along which the value of c_2 is measured. One may try to search for an explicit expression via a Taylor expansion, say in the form

$$\Delta\varepsilon = c_2 \,, \qquad \Delta c = Ac_2 + Bc_2^2 + Cc_2^3 + \dots \,. \tag{7.59}$$

However, it is possible to consider a finite number of terms in this expansion. Suppose that we consider a sequence of pictures of the parameter plane near the critical point in smaller and smaller scales, namely $c_1 \propto \delta_1^{-m}$ and $c_2 \propto \delta_2^{-m}$. If we neglect the Taylor coefficient at c_2^k, the deflection from the proper coordinate curve will behave as δ_2^{-mk}, the respective contribution to the senior eigenvector in the evolution operator (7.57) will be of order $\delta_2^{-mk}\delta_1^m$. In accordance with our numerical data for the eigenvalues, we have $\delta_2 < \delta_1$ and $\delta_2^2 < \delta_1$, but $\delta_2^k > \delta_1$ for $k \geq 3$. Hence, it is sufficient to retain in (7.59) only linear and quadratic terms.

In the spirit of the above discussion, we can use the following ansatz for the definition of the scaling coordinates near the TCT critical point

$$\varepsilon = \varepsilon_c + c_2 \,, \qquad a = a_c - c_1 + A_q c_2 + B_q c_2^2 \,. \tag{7.60}$$

The coefficients may be found if the torus-collision bifurcation curve is computed: This curve has to coincide with the coordinate curve c_1=0; coefficient A_q is related to the slope of the bifurcation curve at the TCT point in the original coordinates (ε, a), while B_q is related to its curvature. Accurate computations yield for the quadratic map

$$A_q{=}0.3117076\,, \qquad B_q{=}0.2819\,.$$

Analogously, for the circle map we set

$$\varepsilon = \varepsilon_c + c_2\,, \qquad c = c_c + c_1 + A_c c_2 + B_c c_2^2\,, \tag{7.61}$$

where

$$A_c = -0.3121848\,, \qquad B_c = -2.047\,. \tag{7.62}$$

If the dynamical regime we are dealing with, occurs completely in the domain of phase space, in which the RG procedure is valid, then the nature of the dynamics should be similar at (c_1, c_2) and $(c_1/\delta_1,\ c_2/\delta_2)$, and both regimes may differ only by a characteristic time scale, which is larger at the second point by a factor $\tau = 1/\omega$. Moreover, quantitative characteristics of both regimes are expressed one via another by trivial relations. Say, the Lyapunov exponents are connected as $\lambda(\varepsilon', c') \cong \lambda(\varepsilon, c)/\tau$. Certainly, all this is true for the localized attractors before the bifurcation of smooth collision or intermittency birth.

In Fig. 7.22 we present the Lyapunov charts of dynamical regimes for our two models using gray scale coding. The small fragment selected near the TCT critical point with borders going along coordinate lines of the scaling coordinate system is shown separately. A smaller fragment of the last picture is magnified several times by factors δ_1 and δ_2 along the horizontal and vertical axes, respectively, to demonstrate similarity of the observed topography. Depicting the diagrams under subsequent steps of magnification, we change the rule of the gray scale coding according to the rule of renormalization of the Lyapunov exponent. One can observe a nice coincidence of the plots at subsequent stages of the magnification.

Let us summarize what can be said about the arrangement of the dynamical regimes near the TCT point on the basis of the stated scaling properties of the evolution operators. In the top panel of Fig. 7.23 we reproduce a sketch of the parameter plane near the TCT point in scaling coordinates. Other panels present the bifurcation diagrams and Lyapunov exponent plots corresponding to the paths in the parameter plane marked (a) and (b). The most notable feature of the picture is that at the TCT point two bifurcation curves meet. One, which goes along the coordinate line $c_1 = 0$, is the line of smooth tori collision, and another, the top curve border, corresponds to an intermittent transition from localized to delocalized attractors. From the stated scaling law, it is easy to derive a relation

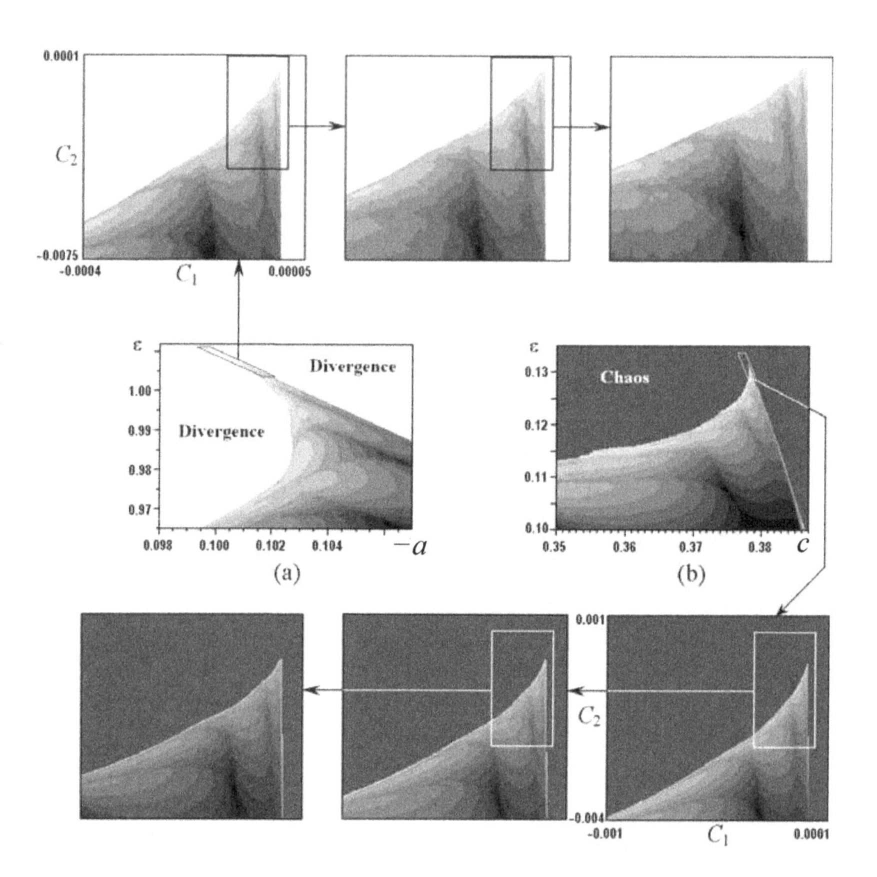

Fig. 7.22 Demonstration of the scaling properties on the parameter plane near the TCT critical points. We plot the Lyapunov exponent for the quadratic map (a) and circle map (b) using a gray scale. For clarity of presentation, only the region of negative exponents is resolved. The main panels (a) and (b) show the Lyapunov charts in the original coordinates on the parameter planes. The insets are depicted in scaling coordinates under successive magnifications: the horizontal scale increases by the factor $\delta_1 \approx 3.65$, and the vertical scale by the factor $\delta_2 \approx 1.81$. The level of the Lyapunov exponent is coded by a gray scale, from light (positive values) to dark (negative values). The coding rule from picture to picture is redefined according to the scaling relation expected for the Lyapunov exponent: the border values for each gray tone are decreased by the factor $\tau = 1.61803$.

for this curve, it must obey the power law

$$c_2 \cong \text{const} \cdot |c_1|^\gamma , \qquad \gamma = \log \delta_2 / \log \delta_1 = 0.470981 \dots . \tag{7.63}$$

Along the road (a), the observed scenario of the onset of complex dynamics involves a bifurcation of smooth collision of stable and unstable tori,

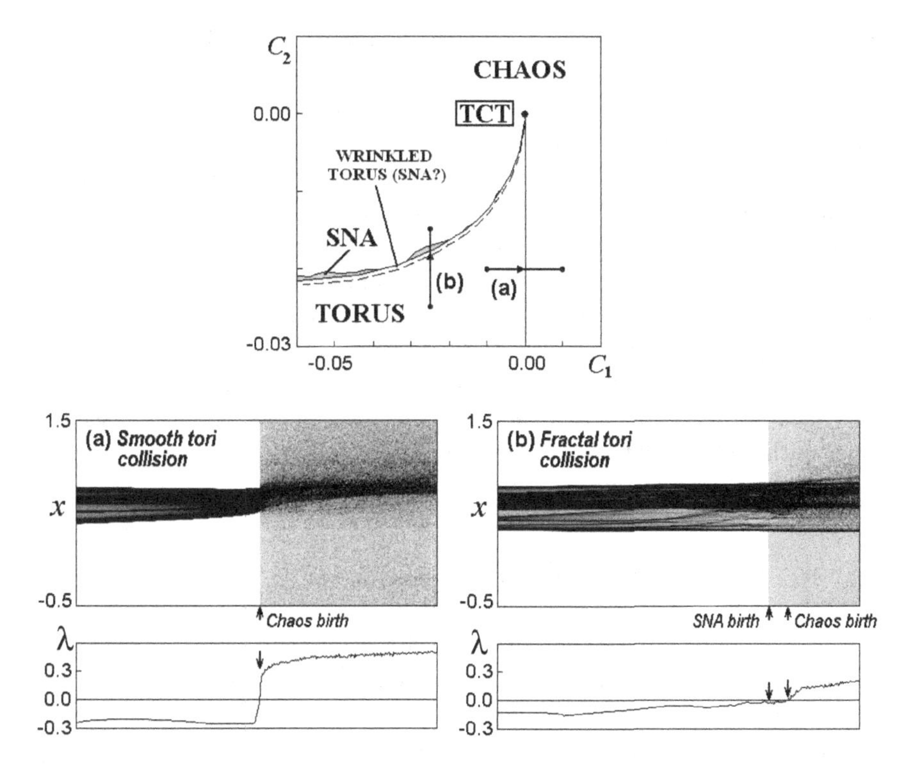

Fig. 7.23 The top panel is a sketch of the parameter plane topography near the TCT point in scaling coordinates in the case of the excluded divergence, as it appears, say, in the driven circle map. The areas of torus, SNA, and chaos are shown. Panels (a) and (b) present the bifurcation diagrams (x variable versus parameter) and Lyapunov exponent plots along the paths in the parameter plane marked with the respective letters. The path (a) corresponds to a collision of smooth stable and unstable tori with their subsequent disappearance and intermittent transition to chaos. Along the path (b) the bifurcation consists in a fractal collision of a smooth unstable torus and an attractor (a wrinkled stable torus or an SNA), with a birth of SNA, which exists in a very narrow strip in the parameter plane; then the transition to chaos follows.

with their coincidence at the bifurcation, and with subsequent disappearance (cf. Section 6.3.1). Beyond that (in the presence of a re-injection mechanism, like in the driven circle map), the result is the onset of an intermittent chaotic regime. On the road (b) the intermittent transition, as observed in computations, occurs at the moment of a touch of the attractor with an unstable invariant set (cf. Section 6.5). At the transition the Lyapunov exponent is negative, thus the only possibility is that the arising intermittent regime corresponds to an SNA. It is not clear whether the localized attractor becomes strange (fractalized) before the intermittent transition takes place, or it remains a smooth (although wrinkled) torus up to the transition point. It is also possible that both alternatives occur at different parts of the bifurcation line. It is clear, however, that the complex form of the attractor at the transition is linked with the fact that the map is non-monotonic in the region of the attractor disposition (the extremum of the map belongs to the domain of the attractor). As to the unstable invariant curve, it is placed completely in the interval of monotonic behavior of the map, and, as seen from computations, it remains smooth up to the moment of bifurcation and further. As can be observed, the regimes of intermittent SNA occupy a very narrow strip in the parameter plane along the chaos border. (At subsequent magnification of the picture in accordance with scaling relations, this strip becomes yet more and more narrow, i.e. its width apparently does not obey the scaling law.)

7.8 RG analysis of the TF critical point

In this section we discuss the criticality of torus fractalization as outlined in Section 7.3.4. Our starting point is system (7.13, 7.14). In the absence of quasiperiodic forcing the map reads $x_{n+1} = x_n/(1 - x_n) + b$. As one increases the parameter b, a tangent bifurcation occurs. The bifurcation consists in a collision of a stable fixed point and an unstable one, with their subsequent disappearance. Beyond this event, a narrow 'channel' remains in the region of former existence of the pair of fixed points, through which the orbits travel for a long time. The closer to the bifurcation point, the larger is the passage time. This picture is a key element of the transition to intermittent dynamics suggested by Pomeau and Manneville [1980b]. Another necessary element of intermittency is the presence of a mechanism of re-injection: after the travel through the channel the orbit should return to repeat this passage again and again.

Adding a quasiperiodic force with the golden mean frequency, we get a map

$$x_{n+1} = x_n/(1 - x_n) + b + \varepsilon \cos 2\pi u_n \ , \qquad u_{n+1} = u_n + \omega \pmod 1 \ . \quad (7.64)$$

In this model, at small amplitudes ε the transition is very similar to the tangent bifurcation, but, instead of the fixed points, the participants are two smooth invariant curves, one stable and another unstable. In the transition, they meet with complete coincidence and annihilate. Then, a region of long-time travel of orbits appears at the place of the former existence of the invariant curves. (Models of the form (7.64), or equivalent, up to a change of variables, were suggested and discussed e.g. by Ketoja and Satija [1997b]; Negi and Ramaswamy [2001a]; Datta et al. [2003].)

At larger ε, the nature of the transition becomes drastically different. At the bifurcation the invariant curves neither coincide, nor remain smooth. What is observed is a formation of a set of narrow sharp 'spikes' on the invariant curves, and mutual touch of them by these spikes. Due to ergodicity of the quasiperiodically forced motion, the spikes, in fact, occupy a dense set on the invariant curve. Obviously, the object we deal with at the bifurcation, is a fractal, and the event is called a non-smooth tori collision (cf. Section 6.3.2).

Both mentioned situations are clearly separated at some critical amplitude of driving; in model (7.64) this separation occurs precisely at $\varepsilon = 2$. This special point on the bifurcation curve of touch of the stable and unstable invariant curves is the *TF critical point* ('torus fractalization') [Kuznetsov 2002a].

To start, let us discuss the procedure of accurate location of the TF critical point.

The first intriguing question is: Why does it correspond exactly to $\varepsilon = 2$? The answer may be obtained in the following way. With a substitution $x_n = 1 - \psi_n/\psi_{n-1}$ the equation (7.64) transforms into the *Harper equation*

$$\psi_{n+1} + \psi_{n-1} + (-2 + b + \varepsilon \cos 2\pi(n\omega + u))\psi_n = 0 , \tag{7.65}$$

known in solid-state physics in the context of wave propagation in a one-dimensional model of a quasicrystal [Harper 1955; Suslov 1982; Ketoja and Satija 1997a]. In this interpretation, ψ_n is a wave function, n is the index on a spatial discrete lattice, ε is the amplitude of a spatially quasiperiodic perturbation, and b is an eigenvalue associated with the frequency, or energy in the quantum-mechanical problem. At small ε, the dominating type of behavior corresponds to propagating waves, with exception of narrow forbidden frequency bands. At large ε, the wavefunctions are localized, thus not propagating. In accordance with the argument of Aubry and Andre [1980], the transition from delocalized to localized states occurs just at $\varepsilon = 2$. Namely, application of the Fourier-like transformation

$$\phi_k = \hat{F}\psi_n = \sum_{n=-\infty}^{n=\infty} \psi_n e^{2\pi i n k \omega} \tag{7.66}$$

to Eq.(7.65) yields a relation of similar form

$$\phi_{k+1} + \phi_{k-1} + (-2 + b' + \varepsilon' \cos 2\pi(n\omega + u))\phi_k = 0 , \tag{7.67}$$

where

$$b' = 2 + 2(b-2)/\varepsilon \ , \ \varepsilon' = 4/\varepsilon \ . \tag{7.68}$$

Localization of the wave function implies delocalization of its transform and vice versa. So, the transition must occur at $\varepsilon = 2$, which corresponds to a fixed point of the equation for ε. It may be proven that the localized solutions of the Harper equation correspond to orbits of the original fractional-linear map which asymptotes at $n \to \infty$ to a localized attractor, and at $n \to -\infty$ to a localized unstable invariant set. Delocalized solutions of the Harper equation correspond to orbits, which depart in the course of motion from the main branch of the map ($x < 1$).

Next, to obtain an accurate estimate for the critical value of b, we may turn to computations based on rational approximations of the frequency parameter by ratios of Fibonacci numbers, $\omega_k = F_{k-1}/F_k$. At frequency ω_k the driving is periodic, and the transition consists in the tangent bifurcation of the period-F_k orbit. It is a collision of a stable cycle with an unstable one, and their Floquet multipliers tend to 1 at the bifurcation. The bifurcation occurs at some b dependent on the initial phase u_0. One can observe that the minimal value of b, at which the bifurcation takes place, corresponds to u_0=0, due to the symmetry of the Harper equation with respect to inversion of u at the origin. Thus we consider a symmetric solution, for which $\psi_1 = \psi_{-1} = \frac{1}{2}(-2 + b + \varepsilon)\psi_0$, and, hence,

$$x_0 = (b + \varepsilon)/(b + \varepsilon - 2) \ . \tag{7.69}$$

In the computations we set $\varepsilon = 2$, $\omega = \omega_k$. Furthermore, we set the initial conditions $x = x_0$ and $u = 0$, and try to adjust the only unknown parameter b to have a cycle of period F_k, i.e., to get $x_{F_k} = x_0$. The results of such computations are presented in Table 7.4. One can observe an evident convergence of the sequence of b values. Thus we conclude that the critical point is located at

$$(\varepsilon, \, b)_{TF} = (2, \, -0.597515185\ldots) \ . \tag{7.70}$$

To find out of what kind is the solution of the RG equation associated with the TF critical point, we may try to compute the functions g_k

Table 7.4 Numerical data for the values of b in the map (7.64) corresponding to the cycle collision at $u=0$ for the critical amplitude $\varepsilon=2$.

ω_k	b
8/13	−0.5989496730498198
13/21	−0.5993564164969890
21/34	−0.5975700101088623
34/55	−0.5977371349948819
55/89	−0.5975077597293093
89/144	−0.5975427315196966
144/233	−0.5975125430279679
233/377	−0.5975187094612125
377/610	−0.5975146415227064
610/987	−0.5975156490874847
987/1597	−0.5975150899617102
1597/2584	−0.5975152478940331
2584/4181	−0.5975151698106684
4181/6765	−0.5975151939749183
6765/10946	−0.5975151829388339
10946/17711	−0.5975151865779841

from direct iterations of the map (7.64) in the same way as in the previously analyzed critical situations. A straightforward attempt fails because the variable x in the original map does not correspond to that in the RG equation. It is necessary to perform preliminary a variable change; an appropriate new coordinate may be defined as the distance from the invariant curve $x=\xi(u)$ at the transition, namely,

$$X \propto x-\xi(u), \quad \xi(u)=x_c+Pu+Qu^2\,. \tag{7.71}$$

Here $x_c=(b_{TF}+2)/b_{TF}=-2.34719526$, as follows from Eq.(7.69) with substitution of $\varepsilon=2$, $b=b_{TF}$. P and Q are coefficients estimated numerically via the first and the second derivative of the invariant curve: $P=(\partial\xi/\partial u)_{u=0}=-5.92667$ and $Q=\frac{1}{2}\left(\partial^2\xi/\partial u^2\right)_{u=0}=210.629$.

Now the procedure consists in the following:

(i) For given X and u define the initial conditions for iterations of the map (7.64): $x=X\alpha^{-m}+x_c+Pu+Qu^2$, $u=U$, α being an empirically selected constant, approximately, 2.89.

(ii) Produce F_m iterations of the map (7.64).

(iii) Return to variable X by the inverse relation $X=\alpha^m\left(x_{F_m}-x_c+Pu_{F_m}+Qu_{F_m}^2\right)$.

The computations show that at m sufficiently large the resulting functions clearly become independent of m, i.e. we deal with a fixed point of the RG equation (7.6), see Fig. 7.24.

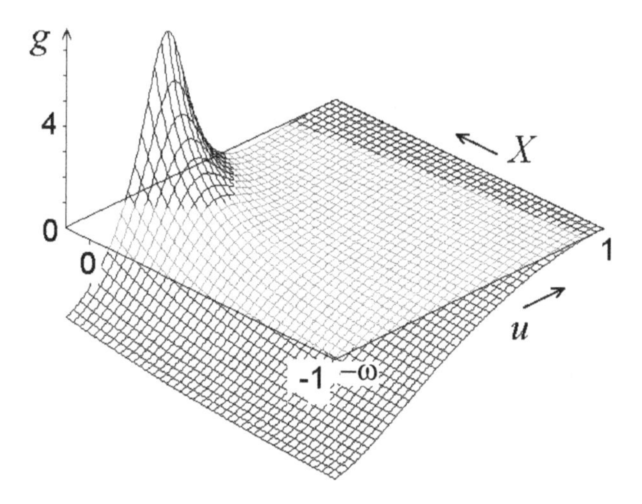

Fig. 7.24 3D plot of the function $g(X, u)$ representing the fixed point of the RG equation associated with the TF critical point.

To continue, we note that due to the fractional-linear nature of the map, the functions obtained at subsequent steps of the RG transformation (7.6) will be fractional-linear, too. The same is true for the limit fixed-point function. Therefore, we may search for solution in the form

$$g(X, u) = (\mathcal{A}(u)X + \mathcal{B}(u))/(\mathcal{C}(u)X + \mathcal{D}(u)) , \tag{7.72}$$

where $\mathcal{A}$, $\mathcal{B}$, $\mathcal{C}$, $\mathcal{D}$ are some functions of u. Without loss of generality, we require them to satisfy additional conditions $\mathcal{A}(u)\mathcal{D}(u) - \mathcal{B}(u)\mathcal{C}(u) \equiv 1$, and $\mathcal{C}(0) = 1$. Substituting (7.72) into (7.6) we arrive at the fixed-point RG equation in terms of the functions $\mathcal{A}$, $\mathcal{B}$, $\mathcal{C}$, $\mathcal{D}$:

$$\begin{bmatrix} \mathcal{A}(u) & \mathcal{B}(u) \\ \mathcal{C}(u) & \mathcal{D}(u) \end{bmatrix} = \begin{bmatrix} \mathcal{A}(\omega^2 u + \omega) & \alpha^2 \mathcal{B}(\omega^2 u + \omega) \\ \alpha^{-2}\mathcal{C}(\omega^2 u + \omega) & \mathcal{D}(\omega^2 u + \omega) \end{bmatrix} \cdot \begin{bmatrix} \mathcal{A}(-\omega u) & \alpha \mathcal{B}(-\omega u) \\ \alpha^{-1}\mathcal{C}(-\omega u) & \mathcal{D}(-\omega u) \end{bmatrix} . \tag{7.73}$$

The solution has been found numerically, and the coefficients of polynomial expansions for $\mathcal{A}(u)$, $\mathcal{B}(u)$, $\mathcal{C}(u)$, $\mathcal{D}(u)$ are listed in [Kuznetsov 2002a] (see also the data on the web page http://www.sgtnd.narod.ru/science/alphabet/eng/goldmean/tf.htm). The factor α was also computed, the results are

$$\alpha = 2.890053525\ldots \qquad \text{and} \qquad \beta = -\omega^{-1} = 1.6180339\ldots . \tag{7.74}$$

These two constants are responsible for the scaling properties of the critical attractor. In Fig. 7.25 the left panel shows the portrait of the attractor in the conventional (u, x)-plane, it looks like a wrinkled curve. A selected curvilinear quadrilateral area has sides coinciding with coordinate lines of the local scaling coordinate system (7.71). The interior of this quadrilateral area is shown separately in the top right panel in scaling coordinates. One can see a self-similarity of the pictures: each next one, drawn with magnification by factors α and β along the vertical and horizontal axes, respectively, reproduces all fine details of the previous one with an excellent accuracy.

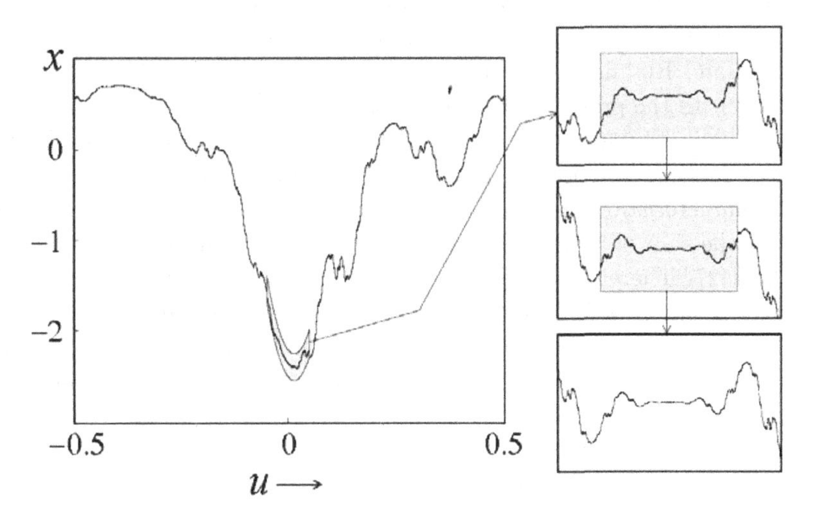

Fig. 7.25 Attractor of the forced fractional-linear map at the TF critical point (the left panel) and illustration of the basic local scaling property: the structure depicted in scaling coordinates reproduces itself under magnification with factors $\alpha = 2.89005$ and $\beta = -1.61803$ along the vertical and the horizontal axes, respectively.

The numerical solution of the eigenvalue problem (7.8) for the fractional-linear fixed point yields two relevant eigenvalues

$$\delta_1 = 3.134272989\ldots, \qquad \delta_2 = 1.618033979\ldots . \tag{7.75}$$

They determine the scaling properties of the evolution operators near the critical point. If we depart from the critical point along the bifurcation curve, the first eigenvector does not contribute; the relevant perturbations are associated with δ_2. If we choose a transversal direction, say, along the axis b, a perturbation of the first kind appears.

In the case under consideration we have $\delta_1 > \delta_2$ and $\delta_1 > \delta_2^2$, but $\delta_1 < \delta_2^3$, so quadratic terms must be taken into account in the parameter change. The scaling coordinates (C_1, C_2) are related to the parameters of the original map as

$$b = b_{TF} + C_1 - 0.64938C_2 - 0.33692C_2^2 \,, \qquad \varepsilon = 2 + C_2 \,. \tag{7.76}$$

To examine the scaling associated with the nontrivial constant δ_1, let us consider the durations of laminar phases in the intermittent dynamics generated by the map with an added branch ensuring the presence of a reinjection mechanism ((7.13, 7.14)). In the usual Pomeau – Manneville intermittency of type I the average duration of the laminar phases behaves as $\langle t_{lam} \rangle \propto \Delta b^\nu$ with $\nu = 0.5$ [Pomeau and Manneville 1980b; Hirsch et al. 1982; Hu and Rudnick 1982] (here Δb is a deviation of parameter b from criticality). In the presence of the quasiperiodic force the same law is valid in the subcritical region, $\varepsilon < 2$. In the critical case $\varepsilon = 2$ the exponent is distinct. Indeed, as it follows from the RG results, to observe an increase of the characteristic time scale by factor $\tau = \omega^{-1} = 1.61803$ we have to decrease the shift of parameter b from the bifurcation threshold by a factor $\delta_1 = 3.13427$. Therefore the exponent must be $\nu = \log \tau / \log \delta_1 \cong 0.42123$. Figure 7.26 shows data of numerical experiments with the fractional-linear map. At each fixed ε an average duration of the passage through the 'channel' near the formerly existing attractor-repeller pair was computed depending on Δb for an ensemble of orbits with random initial conditions. The results are plotted in a double logarithmic scale. For particular values $\varepsilon = 1.7$ (subcritical) and 2 (critical) the dependencies fit straight lines of a definite slope, estimated as 0.508 and 0.424, in a good agreement with the theory. At subcritical ε slightly less than 2 one can observe a 'crossover' phenomenon: the slope changes from the critical to the subcritical value at some intermediate value of Δb.

In the top panel of Fig. 7.27 we reproduce a sketch of the parameter plane near the TF point in scaling coordinates for the model map (7.13). Other panels marked (a) and (b) present the bifurcation diagrams and Lyapunov exponent plots corresponding to the paths in the parameter plane designated with respective letters. The point TF is located on the border of stability of a torus-attractor and separates the situations of smooth and non-smooth tori collisions. Note the differences with the TCT point dis-

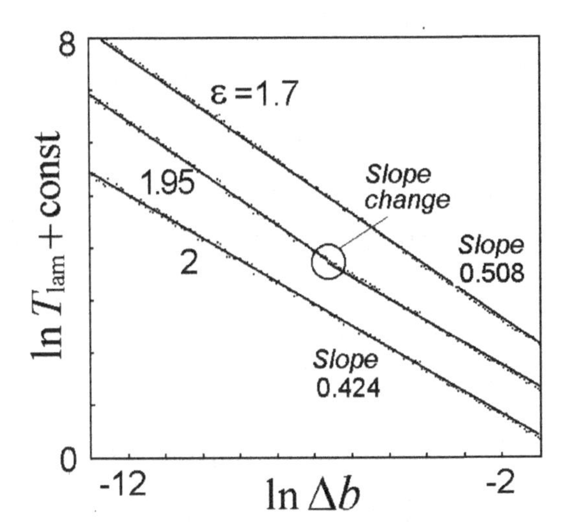

Fig. 7.26 Data of numerical experiments with the fractional-linear map: average duration of passage through the 'channel' versus distance from the bifurcation threshold for three values of ε in double logarithmic scale. Observe the 'crossover' phenomenon, the slope changes from critical to subcritical value at some intermediate value of Δb for $\varepsilon = 1.95$.

cussed in the previous section. First, there is no break of the bifurcation curve; the point TF is a special point of a *smooth* bifurcation curve (in scaling coordinates it looks as a vertical straight line). Second, the non-smooth tori collision is of a special type: both two partners (the stable and the unstable invariant curve) become fractal at the collision: on each of them narrow sharp 'spikes' are formed; at the collision they touch with these spikes.

Crossing the bifurcation curve gives rise to an intermittent dynamics, provided a re-injection mechanism is present. At the part of the bifurcation curve corresponding to a smooth tori collision the intermittency is analogous to that studied by Pomeau and Manneville associated with a tangent bifurcation of one-dimensional maps (in particular, the critical indices are the same). At the part of the border corresponding to the non-smooth tori collision the intermittency gives rise to SNA.

From the point of view of the theory of one-dimensional maps, it is not surprising that the TCT and TF point belong to distinct universality classes: The first occurs in a situation of negative Schwarzian derivative, and the second – of zero Schwarzian derivative. (The Schwarzian derivative for a function $f(x)$ is defined as $S = f'''/f' - \frac{3}{2}(f''/f')^2$, and the main

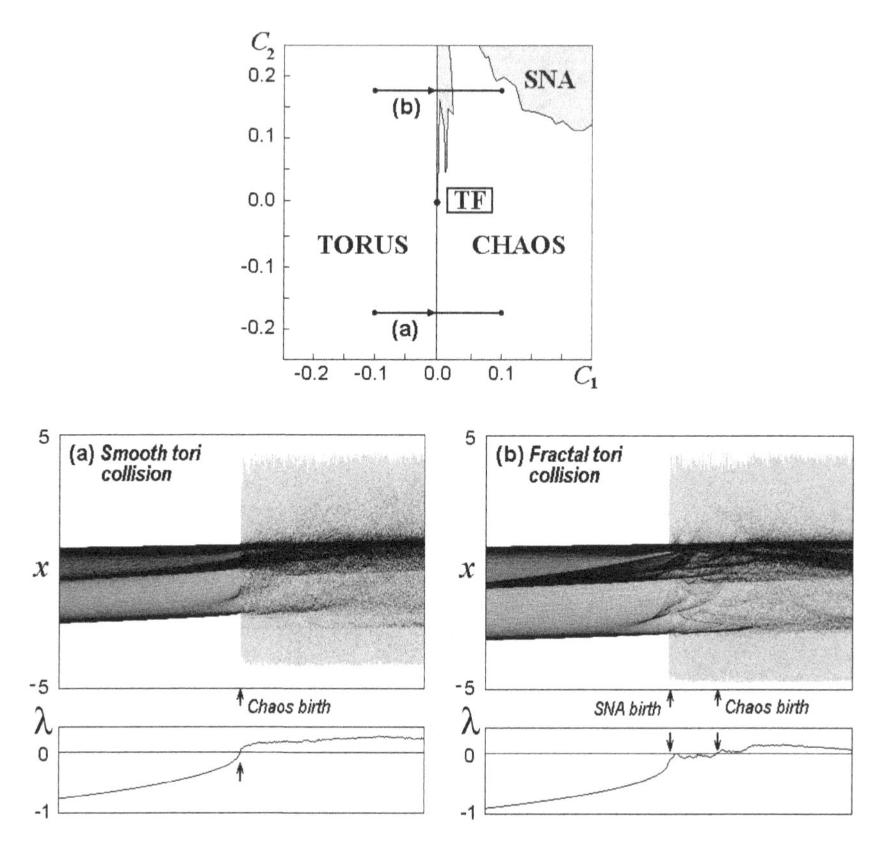

Fig. 7.27 The top panel is a sketch of the parameter plane topography near the TF point in scaling coordinates in the case of presence of the re-injection mechanism. The areas of torus, SNA, and chaos are shown. Panels (a) and (b) present the bifurcation diagrams (x variable versus parameter) and Lyapunov exponent plots along the paths in the parameter plane marked with the respective letters. The path (a) corresponds to a collision of smooth stable and unstable tori with their subsequent disappearance and intermittent transition to chaos. On the path (b) the bifurcation consists in a non-smooth collision of a stable and an unstable invariant curve with sharp narrow 'spikes', which appear approaching the bifurcation.

property is that a composition of functions with the same sign of S has a Schwarzian derivative of the same sign [Singer 1978].)

7.9 Critical behavior in realistic systems

Above in this chapter we have dealt with artificial models, driven one-dimensional maps. However, one can expect that the same types of critical behavior will occur in many realistic systems under quasiperiodic driving, e.g. described by higher dimensional dissipative maps or differential equations. Each type of critical behavior gives rise to a certain universality class. The scaling regularities with definite scaling factors, the universal structure of the critical attractor, and the universal arrangement of the structure in a parameter space near the critical point, with a certain collection of transition scenarios, should be present as the attributes of the universality class. Heuristically, the hypothesis of universality follows from the RG argumentation, in the same spirit as the Feigenbaum universality for the classic period doubling: The form of the evolution operators for large time intervals is determined by the structure of the RG equation rather than by a concrete original system. However, formally speaking, this assertion needs a rigorous proof, which is nontrivial for systems distinct from forced one-dimensional maps. Nevertheless, we may suggest some qualitative recommendations for a search of the critical points in realistic systems, and present some examples.

Concerning the critical points of TDT and TCT type, it is clear that they may be found is many nonlinear dissipative systems, which demonstrate, as autonomous ones, the Feigenbaum period-doubling transition to chaos. In the presence of an additional periodic force with a frequency of incommensurate (e.g., the golden mean) ratio with the characteristic frequency of the system, instead of the period-doubling bifurcations, we will have torus-doubling bifurcations terminated at some amplitude of the additional force at the TDT points. Instead of tangent bifurcations, we will get torus collision bifurcation terminated at some amplitude of driving, at the TCT points.

An experimental example has been described in [Bezruchko et al. 2000]. The underlying system is a driven electronic oscillator, the RL-diode circuit (Fig. 7.28). As is known since the early 80-s, under an increase of the forcing amplitude of one-frequency driving, this system demonstrates the Feigenbaum period-doubling cascade [Testa et al. 1982]. Now, we may add an additional frequency component having the golden-mean ratio with the main frequency. Then, on the parameter plane of amplitudes of the two components we expect to observe qualitatively the same picture as in a driven quadratic map: The amplitude of the main driving A_1 at frequency

ω_1 is analogous to parameter a, and the amplitude of the second component A_2 at frequency $\omega_2 = \frac{1}{2}(\sqrt{5}-1)\omega_1$ – to parameter ε.

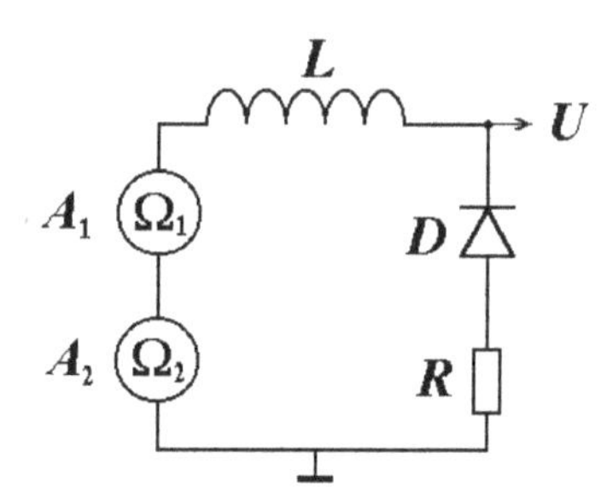

Fig. 7.28 Scheme of the nonlinear electronic circuit under quasiperiodic external driving studied in experiment [Bezruchko et al. 2000]. Two generators produce alternating voltages of amplitudes A_1 and A_2, and frequencies Ω_1 and Ω_2 with a ratio normally equal to the inverse of the golden mean. The values of A_1 and A_2 are regarded as main control parameters of the system.

An observation of phase portraits on the screen of the oscilloscope gave the possibility to find bifurcations of torus doubling. Stroboscopic Poincaré sections of the attractor could be observed using short-time pulse control of the oscilloscope beam, so that the images of points appeared on the screen within time intervals of $2\pi/\omega_1$. Then, the torus is represented by a smooth closed curve, and the doubled torus – by two smooth closed curves, and so forth. Loss of smoothness of the observed curves, or their smearing, could be associated with the onset of SNA or chaos.

To distinguish more carefully regimes of the smooth torus and of SNA an experimental version of the criterion suggested in Chapter 3 was adopted. Namely, using operations of multiplication and division of the basic frequency ω_1, a frequency ratio was selected equal to a rational approximant of the golden mean, namely, 13/21. Then the presence or absence of bifurcations in the system depending on the relative phase between the two components of the external driving was checked. The presence of the bifurcations was regarded as an indication of SNA.

In Fig. 7.29 two charts of dynamical regimes are shown, one obtained directly from the experiment and another computed for the forced logistic map. For the first case, the diagram is drawn on the plane of two amplitudes of the external signal (A_1, A_2), for the second case—on the plane (a, ε). Domains of distinct regimes are indicated by gray levels, and marked with respective inscriptions.

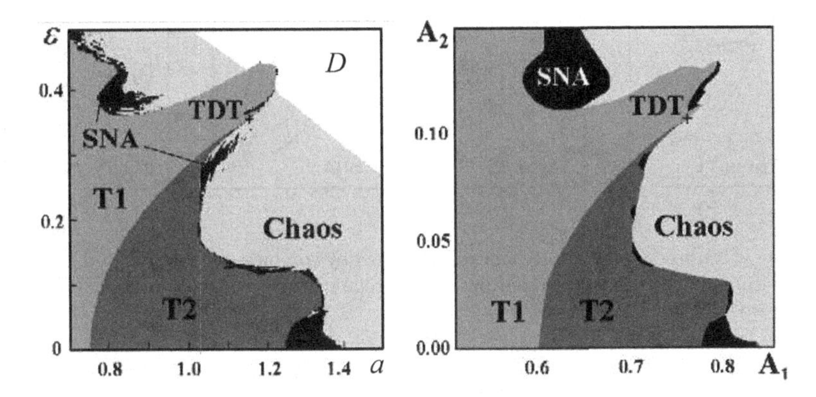

Fig. 7.29 Charts of dynamical regimes computed for the forced quadratic map (left panel) and obtained in the experiment (right panel). Areas of existence of smooth tori T1, T2, T4, arisen one from another via torus-doubling bifurcations, and those of SNA and chaos are marked by respective inscriptions. D is the domain of divergence of iterations for the quadratic map. The TDT critical point is indicated by a little cross. For the experimental system amplitudes of two components of the external force, A_1 and A_2 are plotted along the coordinate axes, for the quadratic map coordinates are the control parameter a and the amplitude of the external force ε.

Using an analog-to-digital converter, the output signal of the system, proportional to the voltage on the resistor U, could be stored in a computer and processed. In particular, using data for the output voltage with a time step equal to one period of the basic frequency $2\pi/\omega_1$, we plot them in coordinates (U_n, U_{n+1}) to obtain the iteration diagrams. In Fig. 7.30 some examples of such diagrams are presented, corresponding to the regimes of torus, doubled torus, SNA, and chaos. Both charts in Fig. 7.29 look remarkably similar. In particular, one can see there the torus-doubling bifurcation line, which separates regions of torus T1 and doubled torus T2. In the experiment, by tuning simultaneously two parameters A_2 and A_1, it was sufficiently easy to move along the bifurcation line and to find its terminal point, i.e., the TDT critical point. Fig. 7.31 shows how the iteration diagrams of the dynamics evolve in this process. Observe that at the TDT point the attractor takes a very specific form, similar both in the experimental system and in the quadratic map: It has a fractal-like shape with a break at the most left point.

Thus, the experiment confirms, at least on a qualitative level, the dynamical behavior associated with the existence of the TDT critical point, the parameter plane arrangement near this point, and the main peculiarities of the dynamics. Apparently, this experimental system indeed relates to

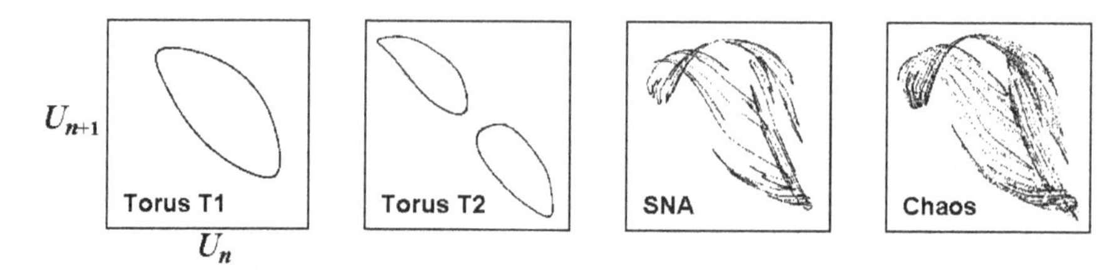

Fig. 7.30 Examples of iteration diagrams obtained by computer processing of data from the experimental system. The diagrams shown relate to torus, double torus, SNA, and chaos, respectively.

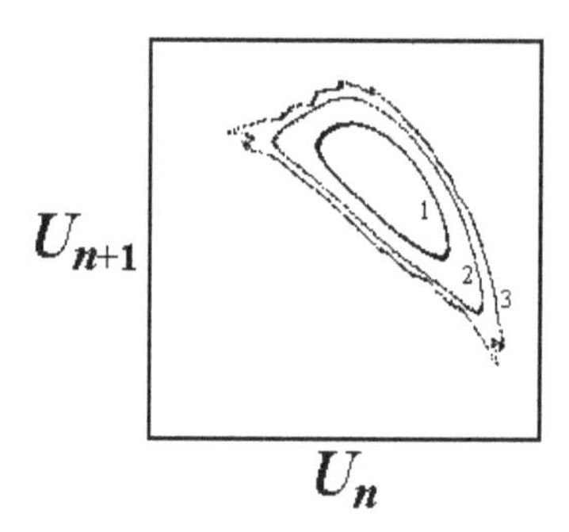

Fig. 7.31 Evolution of the iteration diagram while moving along the torus-doubling bifurcation curve towards the TDT critical point in the experimental system, the quasiperiodically driven RL-diode circuit. Observe the fractal-like shape of the curve representing the critical attractor.

the same universality class as the quasiperiodically forced quadratic map.

Suppose we consider the synchronization of a self-oscillating system with periodic pulses near the border of the mode locking. In presence of an additional frequency component, say, of modulation of the pulse intensity with an incommensurate frequency, the critical situations of TCT and TF type may occur at a sufficiently large amplitude of the additional frequency component. Indeed, in some approximation the problem mentioned reduces to the forced circle map. The appearance of the TF critical behavior in subcritical and critical situations, and of TCT in a supercritical situation was mentioned in the previous sections in this Chapter.

The TF critical point is expected to be very typical in many nonlinear dissipative systems under a quasiperiodic driving, which demonstrate (as autonomous) the saddle-node bifurcation. In particular, it was observed numerically in a model of quasiperiodically forced Josephson junction [Kuznetsov and Neumann 2003]. The model relates to the highly dissipative case and is governed by the ordinary differential equation of first order, represented in dimensionless variables as

$$\dot{x} = -\cos x + I + b\cos t + \varepsilon\cos(\omega t + \varphi_0)\,, \tag{7.77}$$

where $x = \phi - \pi/2$, and ϕ is the Josephson phase – the phase difference of the collective wave functions of the Cooper pairs in two pieces of the superconductor constituting the junction, I is the constant component of current through the junction, b and ε are amplitudes of two alternate components of the current with frequency ratio $\omega = \frac{1}{2}(\sqrt{5} - 1)$, φ_0 – a constant initial phase for the second frequency component. (The same equation is appropriate for a mechanical system, a pendulum placed in a viscous medium, with supplied constant angular momentum I and driven by a two-frequency force.)

In Fig. 7.32 we reproduce a chart of dynamical regimes, where the bottom (white) area is a region of stability of the torus, corresponding to oscillations of the phase φ in a restricted interval. The critical point is placed on the border of this stability region and separates two distinct bifurcation situations, the smooth and fractal tori collision, see illustrations with phase portraits in Fig. 7.33. A feature of this system is that chaos does not arise; the smooth tori collision gives birth to a three-frequency quasiperiodic motion. The non-smooth tori collision gives birth to an SNA with a subsequent transition to three-frequency quasiperiodicity.

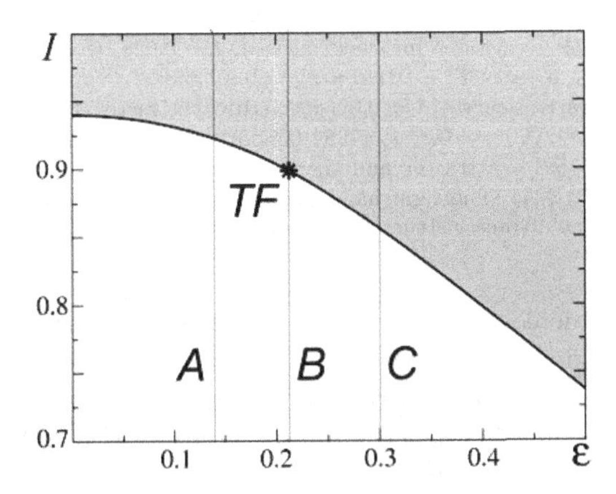

Fig. 7.32 Parameter plane chart of Eq. (7.77) at $b = 0.5$, the regions A, B, C are explained in the caption of Fig. 7.33.

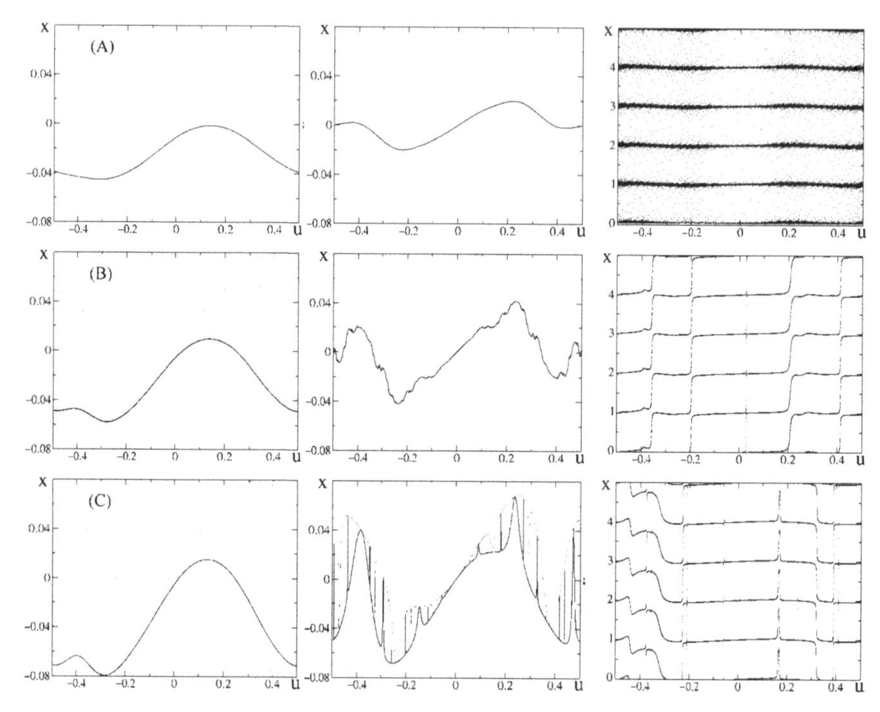

Fig. 7.33 Phase portraits on the plane of variables u and x (panels A, B, C) at representative points of the vertical lines of Fig.7.32 marked with the respective letters. The phase portraits are plotted with the technique of stroboscopic Poincaré cross-sections $t = 0\,(\text{mod } 2\pi)$. Attractors are depicted by solid, and unstable invariant sets by dashed lines: (A) $\varepsilon = 0.14$: $I = 0.9$ (smooth attractor and the unstable invariant curve), $I = 0.922524$ (a single semistable invariant curves), $I = 0.928$ (the attractor is a three-frequency motion on a torus T^3); (B) $\varepsilon = \varepsilon_c = 0.217358$: $I = 0.88$ (smooth attractor and the unstable invariant curve), $I = 0.896801$ (critical attractor, the wrinkled invariant curve), $I = 0.9$ (SNA); (C) $\varepsilon = 0.3$: $I = 0.84$ (smooth attractor and the unstable invariant curve), $I = 0.855914$ (attractor and the unstable invariant set at the non-smooth collision), $I = 0.87$ (SNA). The right-hand pictures are plotted with variable x defined modulo 5 to show the attractor structure more clearly.

A convincing demonstration that the criticality is related to the TF universality class is given in Fig. 7.34. This is a portrait of the critical attractor in the Poincaré section of the extended phase space using surfaces of constant phase of the first frequency component, $t = 0\,(\text{ mod } 2\pi)$, on the plane of variables $u = (\omega t + \varphi_0)/2\pi$ and x. Depicted in properly selected scaling coordinates it demonstrates self-similarity: The picture reproduces itself under magnification with factors $\alpha = 2.890053$ and $\beta = -\omega^{-1}$=1.618034 along two coordinate axes. Moreover, the shape of the object is in a good correspondence with that for a model map used for the study of the TF criticality (see Fig. 7.25).

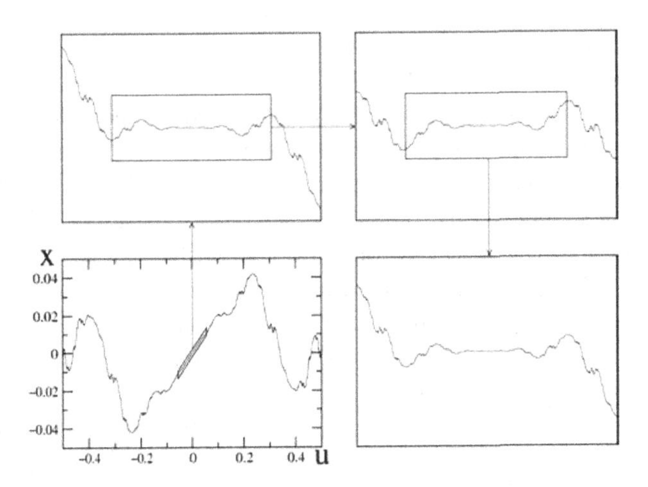

Fig. 7.34 Attractor at the critical point in natural coordinates (u, x) in the Poincaré section and illustration of its self-similarity in scaling coordinates. Factors of magnification are $\alpha = 2.89005$ and $\beta = 1/\omega = 1.618034$ for the vertical and horizontal axes, respectively.

7.10 Conclusion

Renormalization group analysis, which is of principal significance in problems concerning the onset of chaos, also appears to be a valuable instrument for a study of the transitions in quasiperiodically forced systems involving strange nonchaotic attractors. In this Chapter we have formulated an RG equation that allows analyzing a number of types of critical behavior in quasiperiodically driven systems with the golden-mean frequency ratio.

Due to the fact that the evolution operators for large time intervals are determined by the structure of the RG transformation rather than by the concrete form of the original system, each of the examined types of critical behavior is associated with a certain universality class and may occur in systems of different nature.

An analogous situation takes place in the phase transition theory, where the concepts of RG, universality and scaling are borrowed from. Due to a universal character of regularities intrinsic to the critical behavior of matter at the phase transitions, it appears productive to use simple models, constructed with a very rough account of the inter-atomic interactions, but representing a universality class of interest (e.g. the models of Ising, Heisenberg etc.). In the same way, the RG analysis presents a methodological basis for application of simple models in nonlinear dynamics, and for the establishment and analysis of fundamental quantitative regularities intrinsic to the behavior of nonlinear systems at the border of complex dynamics, like chaos or SNA. The construction of such models must be recognized as a self-contained significant task of the theory.

The simple models represented by driven one-dimensional maps studied in this Chapter – the pitchfork bifurcation model, quadratic map, circle map, fractional-linear map in vicinities of their critical points are representatives of the respective universality classes, and may serve as basic examples for the consideration of fundamental aspects of the dynamics at the transitions and near them.

We have examined several critical situations intrinsic to the quasiperiodically driven systems with the golden-mean frequency ratio: the blowout transition, the torus doubling terminal point TDT, the torus collision terminal point TCT, the torus fractalization at the intermittency threshold TF. For each of them, a certain type of the RG equation solution is established, the numerical solution for the functions determining the long-term evolution operators are obtained, the universal constants responsible for the phase space and parameter space scaling properties are accurately estimated. An important general conclusion is that the critical points play a role of, say, organizing centers for the parameter space arrangement. Any neighborhood of a critical point contains all distinct dynamical regimes relevant for the system. Indeed, due to the scaling properties the picture of dynamical regimes is self-similar, reproducing *ad infinitum* in smaller and smaller scales. In other words, local analysis of the parameter space structure contains in a concentrated form information of possible types of dynamical behavior.

In Fig. 7.35 we reproduce schematic pictures of the universal topography of the parameter plane near the critical points TDT, TCT, and TF. In particular, in a vicinity of the TDT point one can observe scenarios of transitions to SNA and chaos like the birth of SNA due to collision of a doubled torus with their unstable parent torus, the intermittent transition, and the torus fractalization. At the TCT point two bifurcation lines of the intermittent transition meet; one corresponds to the birth of intermittent chaos after a smooth tori collision, and another to an intermittent transition involving SNA as an intermediate type of dynamics. The TF point is a special point on the bifurcation line separating two distinct types of the intermittent transition, the smooth tori collision (with transition to chaos, or to a higher dimensional torus), and the birth of SNA via the non-smooth tori collision.

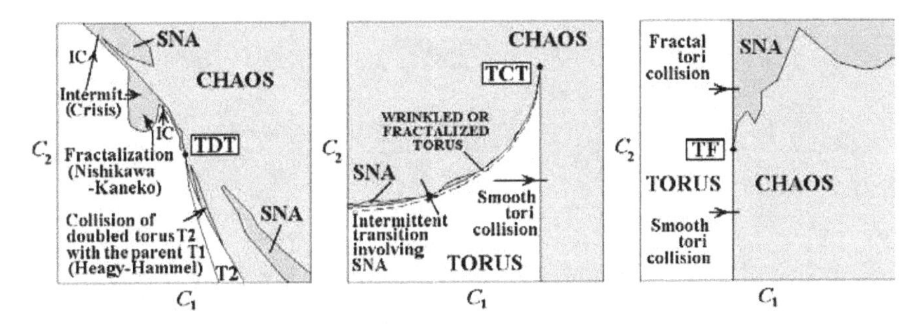

Fig. 7.35 Schematic pictures of the universal topography of the parameter plane near the critical points TDT, TCT, and TF, with indication of the observable transition scenarios. Note that these structures are reproduced in arbitrarily small vicinities of the critical points under appropriate scale change along the vertical and horizontal axes by the universal constants specific for each type of criticality.

At the moment, the number of realistic systems demonstrating the mentioned types of critical behavior is rather small, and a search of further examples is an interesting direction of research. Also, it would be interesting to find out other critical situations, if they exist. Finally, we emphasize again that our consideration on the basis of the RG approach relates only to the case of the golden-mean quasiperiodicity, and generalizations to other frequency ratios, and for quasiperiodicity with a number of incommensurate frequencies more than two, would be desirable and interesting.

7.11 Bibliographic notes

Our presentation here is based on papers [Kuznetsov et al. 1995; Bezruchko et al. 1997; Kuznetsov et al. 1998, 2000; Bezruchko et al. 2000; Kuznetsov 2002b,a; Kuznetsov and Neumann 2003; Kuznetsov 2003]. Some of the renormalization schemes have been treated on a more rigorous basis in [Mestel et al. 2000; Mestel and Osbaldestin 2002; Osbaldestin and Mestel 2003].

Bibliography

V. S. Afraimovich and L. P. Shilnikov. Invariant two-dimensional tori, their destroying and stochasticity. In *Methods of Qualitative Theory of Differential Equations*, pages 3–28. Gorki, 1983. In Russian; English translation Amer. Math. Soc. Transl. (2) **149**, 201–212 (1991).

K. T. Alligood, T. D. Sauer, and J. A. Yorke. *Chaos: An Introduction to Dynamical Systems*. Springer, New York, 1997.

V. S. Anishchenko, T. E. Letchford, and M. A. Safonova. Critical phenomena in the harmonic modulation of two-frequency self-excited oscillations: Transitions to chaos through a three-torus. *Soviet Technical Physics Letters*, 11:223–225, 1985.

V. S. Anishchenko, M. A. Safonova, U. Feudel, and J. Kurths. Bifurcations and transition to chaos through three-dimensional tori. *Int. J. Bif. Chaos*, 4(3):595–607, 1994.

V. S. Anishchenko, T. E. Vadivasova, and O. Sosnovtseva. Mechanisms of ergodic torus destruction and appearance of strange nonchaotic attractors. *Phys. Rev. E*, 53(5):4451–4456, 1996.

J. Argyris, G. Faust, and M. Haase. *An Exploration of Chaos*. North-Holland, Amsterdam, 1994.

P. Ashwin, J. Buescu, and I. Stewart. Bubbling of attractors and synchronization of chaotic oscillators. *Phys. Lett. A*, 193:126–139, 1994.

S. Aubry and G. Andre. Analyticity breaking and Anderson localization in incommensurate lattices. *Annals Israel Phys. Soc.*, 3:133, 1980.

S. Aubry, C. Godrèche, and J. M. Luck. Scaling properties of a structure intermediate between quasiperiodic and random. *J. Stat. Phys.*, 51(5/6): 1033–1075, 1988.

R. Badii and P. F. Meier. Comment on "Chaotic Rabi oscillations under quasiperiodic perturbation". *Phys. Rev. Lett.*, 58(10):1045, 1987.

R. Balescu. *Equilibrium and Nonequilibriun Statistical Mechanics.* Wiley, New York, 1975.

Z. I. Bezhaeva and V. I. Oseledets. An example of a strange nonchaotic attractor. *Functional Analysis and its Applications*, 30(4):223–229, 1996.

B. Bezruchko, S. Kuznetsov, A. Pikovsky, Ye. Seleznev, and U. Feudel. On the dynamics of nonlinear systems under external quasiperiodic force near the terminal point of the torus-doubling bifurcation curve. *Applied Nonlinear Dynamics*, 5(6):3–19, 1997.

B. P. Bezruchko, S. P. Kuznetsov, and Ye. P. Seleznev. Experimental observation of dynamics near the torus-doubling terminal critical point. *Phys. Rev. E*, 62(6):7828–7830, 2000.

P. M. Blekher, H. R. Jauslin, and J. L. Lebowitz. Floquet spectrum for two-level systems in quasiperiodic time-dependent fields. *J. Stat. Phys.*, 68(1/2):271–310, 1992.

A. Bondeson, E. Ott, and T. M. Antonsen. Quasiperiodically forced damped pendula and Schrödinger equation with quasiperiodic potentials: implication of their equivalence. *Phys. Rev. Lett.*, 55(20):2103–2106, 1985.

H. Broer, G. B. Huitema, F. Takens, and B. L. J. Braaksma. Unfoldings and bifurcations of quasi-periodic tori. *Mem. Amer. Math. Soc.*, 83(421): 1–175, 1990.

C. Chandre, H. R. Jauslin, G. Benfatto, and A. Celletti. Approximate renormalization-group transformation for Hamiltonian systems with three degrees of freedom. *Phys. Rev. E*, 60(5):5412–5421, 1999.

P. R. Chastell, P. A. Glendinning, and J. Stark. Locating bifurcations in quasiperiodically forced systems. *Phys. Lett. A*, 200(1):17–26, 1995.

I. P. Cornfeld, S. V. Fomin, and Ya. G. Sinai. *Ergodic Theory.* Springer, New York, 1982.

A. Crisanti, M. Falcioni, G. Paladin, M. Serva, and A. Vulpiani. Complexity in quantum systems. *Phys. Rev. E*, 50(1):138–144, 1994.

P. Cvitanović. *Universality in chaos.* Adam Hilger, Bristol, 1989.

S. Datta, S. Negi, R. Ramaswamy, and U. Feudel. Critical strange nonchaotic dynamics in the Fibonacci map. *Int. J. Bif. Chaos*, 15(4):1493–1501, 2005.

S. Datta, R. Ramaswamy, and A. Prasad. Fractalization route to strange nonchaotic dynamics. *Phys. Rev. E*, 70:046203, 2004.

S. Datta, A. Sharma, and R. Ramaswamy. Thermodynamics of critical strange nonchaotic attractors. *Phys. Rev. E*, 68:036104, 2003.

R. L. Devaney. *An Introduction to Chaotic Dynamical Systems.* Addison-

Wesley, Reading Mass., 1989.
M. Ding, C. Grebogi, and E. Ott. Dimensions of strange nonchaotic attractors. *Phys. Lett. A*, 137(4-5):167–172, 1989a.
M. Ding, C. Grebogi, and E. Ott. Evolution of attractors in quasiperiodically forced systems: From quasiperiodic to strange nonchaotic to chaotic. *Phys.Rev. A*, 39(5):2593–2598, 1989b.
W. L. Ditto, S. Rauseo, R. Cawley, C. Gerbogi G. H. Hsu, E. Kostelich, E. Ott, H. T. Savage, R. Segnam, M. L. Spano, and J. A. Yorke. Experimental observation of crisis-induced intermittency and its critical exponent. *Phys. Rev. Lett.*, 63:923–926, 1989.
W. L. Ditto, M. L. Spano, H. T. Savage, S. N. Rauseo, J. Heagy, and E. Ott. Experimental observation of a strange nonchaotic attractor. *Phys. Rev. Lett.*, 65:533, 1990.
M. J. Feigenbaum. The universal metric properties of nonlinear transformations. *J. Stat. Phys.*, 21:669–706, 1979.
M. J. Feigenbaum. Universal behavior in nonlinear systems. *Physica D*, 7 (1-3):16–39, 1983.
M. J. Feigenbaum, L. P. Kadanoff, and S. J. Shenker. Quasiperiodicity in dissipative systems: a renormalization group analysis. *Physica D*, 5(2-3): 370–386, 1982a.
M. J. Feigenbaum, L. P Kadanoff, and S. J. Shenker. Quasiperiodicity in dissipative systems. A renormalization group analysis. *Physica D*, 5: 370–386, 1982b.
U. Feudel and C. Grebogi. Why are chaotic attractors rare in multistable systems. *Phys. Rev. Lett.*, 91:134102, 2003.
U. Feudel, C. Grebogi, B. R. Hunt, and J. A. Yorke. Map with more than 100 coexisting low-period periodic attractors. *Phys. Rev. E*, 54(1):71–81, 1996a.
U. Feudel, C. Grebogi, and E. Ott. Phase-locking in quasiperiodically forced systems. *Phys. Reports*, 290:11–25, 1997.
U. Feudel, C. Grebogi, L. Poon, and J. A. Yorke. Dynamical properties of a simple mechanical system with a large number of coexisting periodic attractors. *Chaos Solitons Fractals*, 9(1-2):171–180, 1998a.
U. Feudel, J. Kurths, and A. Pikovsky. Strange nonchaotic attractor in a quasiperiodically forced circle map. *Physica D*, 88(3-4):176–186, 1995a.
U. Feudel, A. Pikovsky, and A. Politi. Renormalization of correlations and spectra of a strange nonchaotic attractor. *J. Phys. A*, 29:5297–5311, 1996b.
U. Feudel, A. S. Pikovsky, and M. A. Zaks. Correlation properties of

quasiperiodically forced two-level system. *Phys. Rev. E*, 51(3):1762–1769, 1995b.

U. Feudel, A. Witt, Y.-C. Lai, and C. Grebogi. Basin bifurcation in quasiperiodically forced systems. *Phys. Rev. E*, 58:3060–3066, 1998b.

W. J. Fu, D. H. He, P. L. Shi, W. Kang, and G. Hu. Scaling of torus-doubling terminal points in a quasi-periodically forced map. *Chinese Physics*, 11(1):17–20, 2002.

H. Fujisaka and T. Yamada. A new intermittency in coupled dynamical systems. *Prog. Theor. Phys.*, 74(4):918–921, 1985.

J. A. C. Gallas, C. Grebogi, and J. A. Yorke. Vertices in parameter space: double crises which destroys chaotic attractors. *Phys. Rev. Lett.*, 71: 1359–1362, 1993.

T. Geisel. Quasiperiodicity versus mixing instability in a kicked quantum system. *Phys. Rev. A*, 41(6):2989–2994, 1990.

P. Glendinning. Intermittency and strange nonchaotic attractors in quasi-periodically forced circle maps. *Phys. Lett. A*, 244(6):545–550, 1998.

P. Glendinning. Non-smooth pitchfork bifurcation. *Discrete and Continuous Dynamical Systems - Series B*, 4(2):457–464, 2004.

P. Glendinning, U. Feudel, A. Pikovsky, and J. Stark. The structure of mode-locking regions in quasi-periodically forced circle maps. *Physica D*, 140(1):227–243, 2000.

P. Glendinning and J. Wiersig. Fine structure of mode-locking regions of the quasiperiodically forced circle map. *Phys. Lett. A*, 257:65–69, 1999.

C. Grebogi, S. McDonald, E. Ott, and J. A. Yorke. Final state sensitivity:an obstruction to predictability. *Phys. Lett. A*, 99(9):415–418, 1983a.

C. Grebogi, H. E. Nusse, E. Ott, and J. A. Yorke. Basic sets: Sets that determine the dimension of basin boundaries. In J. C. Alexander, editor, *Dynamical Systems*, volume 1342 of *Lecture Notes in Mathematics*, pages 220–250. Springer-Verlag, Berlin, 1988.

C. Grebogi, E. Ott, S. Pelikan, and J. A. Yorke. Strange attractors that are not chaotic. *Physica D*, 13(1-2):261–268, 1984.

C. Grebogi, E. Ott, F. Romeiras, and J. A. Yorke. Critical exponents for crisis-induced intermittency. *Phys. Rev. A*, 36(11):5365–5380, 1987a.

C. Grebogi, E. Ott, and J. A. Yorke. Chaotic attractors in crisis. *Phys. Rev. Lett*, 48(22):1507–1510, 1982.

C. Grebogi, E. Ott, and J. A. Yorke. Are three-frequency quasiperiodic orbits to be expected in typical nonlinear dynamical systems? *Phys. Rev. Lett*, 51(5):339–342, 1983b.

C. Grebogi, E. Ott, and J. A. Yorke. Fractal basin boundaries, long-lived chaotic transients, and unstable-unstable pair bifurcation. *Phys. Rev. Lett.*, 50(13):935–938, 1983c.

C. Grebogi, E. Ott, and J. A. Yorke. Attractors on an N-torus. quasiperiodicity versus chaos. *Physica D*, 15:354–373, 1985.

C. Grebogi, E. Ott, and J. A. Yorke. Critical exponent of chaotic transients in nonlinear dynamical systems. *Phys. Rev. Lett.*, 57(11):1284–1287, 1986a.

C. Grebogi, E. Ott, and J. A. Yorke. Metamorphoses of basin boundaries in nonlinear dynamical systems. *Phys. Rev. Lett.*, 56:1011–1014, 1986b.

C. Grebogi, E. Ott, and J. A. Yorke. Basin boundary metamorphoses: Changes in accessible boundary orbits. *Physica D*, 24:243–262, 1987b.

J. M. Greene. A method for determining a stochastic transition. *J. Math. Phys.*, 20:760–768, 1979.

E. G. Gwinn and R. M. Westervelt. Fractal basin boundaries and intermittency in the driven damped pendulum. *Phys. Rev. A*, 33:4143, 1986.

P. G. Harper. Single band motion of conduction electrons in a uniform magnetic field. *Proc. Phys. Soc. A*, 68:874–878, 1955.

J. Heagy and W. L. Ditto. Dynamics of a two-frequency parametrically driven duffing oscillator. *J. Nonlinear Science*, 1(4):423–456, 1991.

J. F. Heagy and S. M. Hammel. The birth of strange nonchaotic attractors. *Physica D*, 70:140–153, 1994.

J. E. Hirsch, M. Nauenberg, and D. J. Scalapino. Intermittency in the presence of noise. A renormalization group formulation. *Phys. Lett. A*, 87:391, 1982.

B. Hu and J. Rudnick. Exact solutions to the Feigenbaum renormalization-group equations for intermittency. *Phys. Rev. Lett.*, 48:1645, 1982.

B. R. Hunt and E. Ott. Fractal properties of robust strange nonchaotic attractors. *Phys. Rev. Lett.*, 87(25):254101, 2001.

M. Iansiti, Qing Hu, R. M. Westervelt, and M. Tinkham. Noise and chaos in a fractal basin boundary regime of a Josephson junction. *Phys. Rev. Lett.*, 55:746–749, 1985.

N. Yu. Ivankov and S. P. Kuznetsov. Complex periodic orbits, renormalization and scaling for quasiperiodic golden-mean transition to chaos. *Phys. Rev. E*, 63(4):046210, 2001.

A. Yu. Jalnine and A. H. Osbaldestin. Smooth and nonsmooth dependence of Lyapunov vectors upon the angle variable on a torus in the context of torus-doubling transitions in the quasiperiodically forced Hénon map. *Phys. Rev. E*, 71:016206, 2005.

K. Kaneko. Fates of three-torus I. Double devil's staircase in lockings. *Prog.*

Theor. Phys., 71(2):282–294, 1984a.

K. Kaneko. Fractalization of torus. *Prog. Theor. Phys.*, 71(5):1112–1114, 1984b.

K. Kaneko. Oscillation and doubling of torus. *Prog. Theor. Phys.*, 72(2): 202–215, 1984c.

K. Kaneko. *Collapse of tori and genesis af chaos in dissipative systems.* World Scientific, Singapore, 1986.

T. Kapitaniak. Generating strange nonchaotic trajectories. *Phys. Rev. E*, 47(2):1408–1410, 1993.

T. Kapitaniak. Distribution of transient Lyapunov exponents of quasiperiodically forced systems. *Prog. Theor. Phys.*, 93(4):831–833, 1995.

T. Kapitaniak, E. Ponce, and J. Wojewoda. Route to chaos via strange non-chaotic attractors. *J. Phys. A: Math., Gen.*, 23(8):L383–L387, 1990.

A. Katok and B. Hasselblatt. *Introduction to the Modern Theory of Dynamical Systems.* Cambridge University Press, 1995.

G. Keller. A note on strange nonchaotic attractors. *Fundamenta Mathematicae*, 151:139–148, 1996.

J. A. Kennedy and J. A. Yorke. Basins of Wada. *Physica D*, 51:213–225, 1991.

J. A. Ketoja and I. I. Satija. Self-similarity and localization. *Phys. Rev. Lett.*, 75(14):2762–2765, 1995.

J. A. Ketoja and I. I. Satija. "Critical" phonons of the supercritical Frenkel-Kontorova model: Renormalization bifurcation diagrams. *Physica D*, 104:239–252, 1997a.

J. A. Ketoja and I. I. Satija. Harper equation, the dissipative standard map and strange nonchaotic attractor: Relationship between an eigenvalue problem and iterated maps. *Physica D*, 109(1-2):70–80, 1997b.

R. Ketzmerick, G. Petschel, and T. Geisel. Slow decay of temporal correlations in quantum systems with Cantor spectra. *Phys. Rev. Lett.*, 69:695, 1992.

A. Ya. Khinchin. *Continued Fractions.* The University of Chicago, Chicago, 1949.

I. A. Khovanov, N. A. Khovanova, V. S. Anishchenko, and P. V. E. McClintock. Sensitivity to initial conditions and the Lyapunov exponent of a quasiperiodic system. *(In Russian). ZhTF*, 70(5):112–114, 2000a.

I. A. Khovanov, N. A. Khovanova, P. V. E. McClintock, and V. S. Anishchenko. The effect of noise on strange nonchaotic attractors. *Phys. Lett. A*, 268:315–322, 2000b.

J.-W. Kim, S.-Y. Kim, B. Hunt, and E. Ott. Fractal properties of robust strange nonchaotic attractors in maps of two or more dimensions. *Phys. Rev. E*, 67:036211, 2003a.

S.-Y. Kim and W. Lim. Mechanism for boundary crises in quasiperiodically forced period-doubling systems. *Phys. Lett. A*, 334:160–168, 2005.

S.-Y. Kim, W. Lim, and E. Ott. Mechanism for the intermittent route to strange nonchaotic attractors. *Phys. Rev. E*, 67:056203, 2003b.

A. P. Kuznetsov, S. P. Kuznetsov, and I. R. Sataev. A variety of period-doubling universality classes in multi-parameter analysis of transition to chaos. *Physica D*, 109:91–112, 1997.

S. Kuznetsov, U. Feudel, and A. Pikovsky. Renormalization group for scaling at the torus-doubling terminal point. *Phys. Rev. E*, 57(2):1585–1590, 1998.

S. Kuznetsov, A. Pikovsky, and U. Feudel. Birth of a strange nonchaotic attractor: A renormalization group analysis. *Phys. Rev. E*, 51(3):R1629–R1632, 1995.

S. P. Kuznetsov. Effect of a periodic external perturbation on a system which exhibits an order-chaos transition through period-doubling bifurcations. *JETP Letters*, 39(3):133–136, 1984.

S. P. Kuznetsov. Torus fractalization and intermittency. *Phys. Rev. E*, 65: 066209, 2002a.

S. P. Kuznetsov. A variety of critical phenomena associated with the golden mean quasiperiodicity. *Applied Nonlinear Dynamics (Saratov)*, 10(3):22–39, 2002b.

S. P. Kuznetsov. Generalization of the Feigenbaum-Kadanoff-Shenker renormalization and critical phenomena associated with the golden mean quasiperiodicity. In A. Pikovsky and Yu. Maistrenko, editors, *Syncronization: Theory and application*, volume 109 of *NATO Science Series. II. Mathematics, Physics and Chemistry*, pages 79–100, Dordrecht, Boston, London, 2003. Kluwer.

S. P. Kuznetsov. Effect of noise on the dynamics at the torus-doubling terminal point in a quadratic map under quasiperiodic driving. *Phys. Rev. E*, 72:026205, 2005.

S. P. Kuznetsov and E. Neumann. Torus fractalization and singularities in the current-voltage characteristics for the quasiperiodically forced Josephson junction. *Europhys. Lett.*, 61(1):20–26, 2003.

S. P. Kuznetsov, E. Neumann, A. Pikovsky, and I. R. Sataev. Critical point of tori collision in quasiperiodically forced systems. *Phys. Rev. E*, 62(2): 1995–2007, 2000.

S. P. Kuznetsov and A. S. Pikovsky. Renormalization group for the response function and spectrum of the period–doubling system. *Phys. Lett. A*, 140 (4):166–172, 1989.

Y.-C. Lai, U. Feudel, and C. Grebogi. Scaling behavior of transition to chaos in quasiperiodically driven dynamical systems. *Phys. Rev. E*, 54 (6):6070–6073, 1996.

Y.-C. Lai and C. Grebogi. Intermingled basins and two-state on-off intermittency. *Phys. Rev. E*, 52:R3313–R3316, 1995.

Y.-C. Lai, C. Grebogi, and J. A. Yorke. Sudden change in the size of chaotic attractors: how does it occur? In J. H. Kim and J. Stringer, editors, *Applied Chaos*, pages 441–455. John Wiley & Sons, 1992.

L. D. Landau and E. M. Lifshitz. *Fluid mechanics.* Pergamon Press, Oxford–New York, 1987.

A. J. Lichtenberg and M. A. Lieberman. *Regular and Chaotic Dynamics.* Springer, New York, 1992.

W. Lim and S.-Y. Kim. Dynamical mechanism for band-merging transitions in quasiperiodically forced systems. *Phys. Lett. A*, 335:383–393, 2005.

J. M. Luck, H. Orland, and U. Smilansky. On the response of a two-level quantum system to a class of time-dependent quasiperiodic perturbations. *J. Stat. Phys.*, 53(3/4):551–564, 1988.

R. S. MacKay. A renormalization approach to invariant circles in area-preserving maps. *Physica D*, 7(1-3):283–300, 1983.

B. Mandelbrot. *Fractals - form, chance, and dimension.* Freeman, San Francisco, 1977.

S. W. McDonald, C. Grebogi, E. Ott, and J. A. Yorke. Fractal basin boundaries. *Physica D*, 17(2):125–153, 1985.

B. D. Mestel and A. H. Osbaldestin. Periodic orbits of renormalization for the correlation function of strange nonchaotic attractor. *Math. Phys. EJ*, 6(5):p29, 2000.

B. D. Mestel and A. H. Osbaldestin. Renormalization analysis of correlation properties in a quasiperiodically forced two-level system. *J.Math.Phys.*, 43(7):3458–3483, 2002.

B. D. Mestel and A. H. Osbaldestin. A garden of orchids: a generalized Harper equation at quadratic irrational frequencies. *J. Phys. A: Math. Gen.*, 37:9071–9086, 2004.

B. D. Mestel, A. H. Osbaldestin, and B. Winn. Golden mean renormalization for the Harper equation: The strong coupling fixed point. *J. Math. Phys.*, 41(12):8304–8330, 2000.

J. Milnor. On the concept of attractor. *Commun. Math. Phys.*, 99:177–195, 1985.

F. C. Moon. Fractal boundary for chaos in a two-state mechanical oscillator. *Phys. Rev. Lett*, 53(10):962–964, 1984.

M. Napiórkowski. Scaling of the uncertainty exponent. *Phys. Rev. A*, 33: 4423, 1986.

S. S. Negi, A. Prasad, and R. Ramaswamy. Bifurcations and transitions in the quasiperiodically driven logistic map. *Physica D*, 145(1-2):1–12, 2000.

S. S. Negi and R. Ramaswamy. Critical states and fractal attractors in fractal tongues: Localization in the Harper map. *Phys. Rev. E*, 64:045204, 2001a.

S. S. Negi and R. Ramaswamy. A plethora of strange nonchaotic attractors. *PRAMANA Journal of Physics*, 56(1):47–56, 2001b.

E. Neumann and A. Pikovsky. Quasiperiodically driven Josephson junctions: strange nonchaotic attractors, symmetries and transport. *Eur. Phys. J. B*, 26(2):219–228, 2002.

E. Neumann, I. Sushko, Yu. Maistrenko, and U. Feudel. Synchronization and desynchronization under the influence of quasiperiodic forcing. *Phys. Rev. E*, 67(2):026202, 2003.

S. E. Newhouse, D. Ruelle, and F. Takens. Occurrence of strange axiom A attractors near quasi-periodic flows on T^m (m = 3 or more). *Commun. Math. Phys.*, 64:35–40, 1978.

T. Nishikawa and K. Kaneko. Fractalization of torus revisited as a strange nonchaotic attrator. *Phys. Rev. E*, 54(6):6114–6124, 1996.

H. E. Nusse and J. A. Yorke. Wada basin boundaries and basin cells. *Physica D*, 90(3):242–261, 1996.

A. H. Osbaldestin and B. D. Mestel. Renormalization in quasiperiodically forced systems. *Fluct. Noise Lett.*, 3(2):L251–L258, 2003.

H. M. Osinga and U. Feudel. Boundary crisis in quasiperiodically forced systems. *Physica D*, 141(1-2):54–64, 2000.

H. M. Osinga, J. Wiersig, P. Glendinning, and U. Feudel. Multistability and nonsmooth bifurcations in quasiperiodically forced circle map. *Int. J. Bif. Chaos*, 11(12):3085–3105, 2001.

E. Ott. *Chaos in Dynamical Systems.* Cambridge Univ. Press, Cambridge, 1992.

E. Ott and J. C. Sommerer. Blowout bifurcations: the occurrence of riddled basins and on-off intermittency. *Phys. Lett. A*, 188:39–47, 1994.

B.-S. Park, C. Grebogi, and Y.-Ch. Lai. Abrupt dimension changes at basin boundary metamorphoses. *Int. J. of Bifurcation and Chaos*, 2(3): 544–541, 1992.

B.-S. Park, C. Grebogi, E. Ott, and J. A. Yorke. Scaling of fractal basin boundaries near intermittency transitions to chaos. *Phys. Rev. A*, 40: 1576, 1989.

A. Pikovsky and U. Feudel. Correlations and spectra of strange nonchaotic attractors. *J. Phys. A: Math., Gen.*, 27(15):5209–5219, 1994.

A. Pikovsky and U. Feudel. Characterizing strange nonchaotic attractors. *CHAOS*, 5(1):253–260, 1995.

A. Pikovsky, M. Rosenblum, and J. Kurths. *Synchronization. A Universal Concept in Nonlinear Sciences.* Cambridge University Press, Cambridge, 2001.

A. Pikovsky, M. Zaks, and J. Kurths. Complexity of quasiperiodically driven spin system. *J. Phys. A*, 29:295–302, 1996.

A. S. Pikovsky. On the interaction of strange attractors. *Z. Physik B*, 55 (2):149–154, 1984.

A. S. Pikovsky and P. Grassberger. Symmetry breaking bifurcation for coupled chaotic attractors. *J. Phys. A: Math., Gen.*, 24(19):4587–4597, 1991.

N. Platt, E. A. Spiegel, and C. Tresser. On-off intermittency: A mechanism for bursting. *Phys. Rev. Lett.*, 70:279–282, 1989.

P. Pokorny, I. Schreiber, and M. Marek. On the route to strangeness without chaos in the quasiperiodically forced van der Pol oscillator. *Chaos Solitons Fractals*, 7(3):409–424, 1996.

Y. Pomeau, B. Dorizzi, and B. Grammaticos. Chaotic Rabi oscillations under quasiperiodic perturbation. *Phys. Rev. Lett.*, 56:681–684, 1986.

Y. Pomeau and P. Manneville. Intermittent transition to turbulence in dissipative dynamical systems. *Commun. Math. Phys.*, 74:189–197, 1980a.

Y. Pomeau and P. Manneville. Intermittent transition to turbulence in dissipative dynamical systems. *Comm. Math. Phys.*, 74(2):189, 1980b.

L. Poon, J. Campos, E. Ott, and C. Grebogi. Wada basin boundaries in chaotic scattering. *Int. J. Bif. Chaos*, 6(2):251–265, 1996.

A. Prasad, B. Biswal, and R. Ramaswamy. Strange nonchaotic attractors in driven excitable systems. *Phys. Rev. E*, 68:037201, 2003.

A. Prasad, V. Mehra, and R. Ramaswamy. Intermittency route to strange nonchaotic attractors. *Phys. Rev. Lett.*, 79(21):4127–4130, 1997.

A. Prasad, V. Mehra, and R. Ramaswamy. Strange nonchaotic attractors in the quasiperiodically forced logistic map. *Phys. Rev. E*, 57(2):1576–1584, 1998.

A. Prasad, S. S. Negi, and R. Ramaswamy. Strange nonchaotic attractors (Rewiew). *Int. J. Bif. Chaos*, 11:291–311, 2001.

A. Prasad and R. Ramaswamy. Characteristic distributions of finite-time Lyapunov exponents. *Phys. Rev. E*, 60(3):2761–2766, 1999.

A. Prasad, R. Ramaswamy, I. I. Satija, and N. Shah. Collision and symmetry breaking in the transition to strange nonchaotic attractors. *Phys. Rev. Lett.*, 83(22):4530–4533, 1999.

R. Ramaswamy. Synchronization of strange nonchaotic attractors. *Phys. Rev. E*, 56(6):7294–7296, 1997.

D. Rand, S. Ostlund, J. Sethna, and E. D. Siggia. Universal transition from quasiperiodicity to chaos in dissipative systems. *Phys. Rev. Lett.*, 49(2): 132–135, 1982.

D. Rand, S. Ostlund, J. Sethna, and E. D. Siggia. Universal properties of the transition from quasi-periodicity to chaos in dissipative systems. *Physica D*, 8:303–342, 1983.

A. M. Rocket and P. Szüsz. *Continued Fractions.* World Scientific, Singapore, 1992.

F. J. Romeiras and E. Ott. Strange nonchaotic attractors of the damped pendulum with quasiperiodic forcing. *Phys. Rev. A*, 35(10):4404–4413, 1987.

D. Ruelle and F. Takens. On the nature of turbulence. *Comm. Math. Phys.*, 20:167–192, 1971.

M. Samuelides, R. Fleckinger, L. Touziller, and J. Bellisard. Instabilities of the quantum rotator and transition in the quasi-energy spectrum. *Europhys. Lett.*, 1(5):203–208, 1986.

E. P. Seleznev and A. M. Zakharevich. Dynamics of a nonlinear oscillator under quasi-periodic drive action. *Technical Physics Letters*, 31(9):725–727, 2005.

S. J. Shenker. Scaling behavior in a map of a circle onto itself:empirical results. *Physica D*, 5(2-3):405–411, 1982.

D. L. Shepelyansky. Some statistical properties of simple classically stochastic quantum systems. *Physica D*, 8(1-2):208–222, 1983.

P. L. Shi, D. H. He, W. Kang, W. J. Fu, and G. Hu. Chaoslike behavior in nonchaotic systems at finite computation precision. *Phys. Rev. E*, 6304 (4):6310–6317, 2001.

D. V. Shirkov, D. I. Kazakov, and A. A. Vladimirov, editors. *Renormalization Group (Proc. of Int. Conf., Dubna).* World Sci., Singapur, 1986.

M. D. Shrimali, A. Prasad, R. Ramaswamy, and U. Feudel. Basin bifurcations in quasiperiodically forced coupled systems. *Phys. Rev. E*, 72(3): 036215, 2005.

J. W. Shuai and D. M. Durand. Strange nonchaotic attractor in low-frequency quasiperiodically driven systems. *Int. J. Bifurcation Chaos*, 10(9):2269–2276, 2000.

J. W. Shuai, J. Lian, P. J. Hahn, and D. M. Durand. Positive Lyapunov exponents calculated from time series of strange nonchaotic-attractors. *Phys. Rev. E*, 6402(2):6220–6224, 2001.

J. W. Shuai and K. W. Wong. Nonchaotic attractors with highly fluctuating finite-time lyapunov exponents. *Phys.Rev. E*, 57(5):5332–5336, 1998.

D. Singer. Stable orbits and bifurcation of maps of the interval. *SIAM J. Appl. Math.*, 35(2):260–267, 1978.

J. C. Sommerer, W. L. Ditto, C. Grebogi, E. Ott, and M. L. Spano. Experimental confirmation of the theory of critical exponents of crises. *Phys. Lett. A*, 153:105, 1991.

O. Sosnovtseva, U. Feudel, A. Pikovsky, and J. Kurths. Multiband strange nonchaotic attractors in quasiperiodically forced systems. *Phys. Lett. A*, 218:255–267, 1996.

O. Sosnovtseva, T. E. Vadivasova, and V. S. Anishchenko. Evolution of complex oscillations in a quasiperiodically forced chain. *Phys.Rev. E*, 57 (1):282–287, 1998.

J. J. Stagliano, J.-M. Wersmger, and E. E. Slaminka. Doubling bifurcations of destroyed t2 tori. *Physica D*, 92:164–177, 1996.

J. Stark. Invariant graphs for forced systems. *Physica D*, 109(1-2):163–179, 1997.

J. Stark, U. Feudel, P. A. Glendinning, and A. Pikovsky. Rotation numbers for quasi-periodically forced monotone circle maps. , *Dynamical Systems*, 17(1):1–28, 2002.

H. B. Stewart, Y. Ueda, C. Grebogi, and J. A. Yorke. Double crises in two-parameter dynamical systems. *Phys. Rev. Lett.*, 75(13):2478–2481, 1995.

R. Sturman. Scaling of intermittent behaviour of a strange nonchaotic attractor. *Phys. Lett. A*, 259:355–365, 1999a.

R. Sturman and J. Stark. Semi-uniform ergodic theorems and application to forced systems. *Nonlinearity*, 13:113–143, 1999.

I. M. Suslov. Localization in one-dimensional incommensurable systems. *Sov. Phys. JETP*, 56:612, 1982.

B. Sutherland. Simple system with quasiperiodic dynamics: a spin in a magnetic field. *Phys. Rev. Letters*, 57(6):770–773, 1986.

J. Testa, J. Perez, and C. Jeffries. Evidence for universal behavior of a driven nonlinear oscillator. *Phys. Rev. Lett*, 48(11):714–717, 1982.

Y. Ueda, S. Yoshida, H. B. Steward, and J. M. T. Thompson. Basin explosions and escape phenomena in the twin-welled Duffing oscillator. *Phil. Trans. Roy. Soc. London*, 332:169–186, 1990.

T. E. Vadivasova, O. Sosnovtseva, and A. G. Balanov. Phase multistability in systems with quasiperiodic external force. *Pis'ma v ZhTF*, 25(2):49–56, 1999.

T. E. Vadivasova, O. V. Sosnovtseva, A. G. Balanov, and V. V. Astakhov. Desynchronization in coupled systems with quasiperiodic driving. *Phys. Rev. E*, 61(4):4618–4621, 2000.

A. Vasylenko, Y. Maistrenko, O. Feely, and U. Feudel. Mode-locking in quasi-periodically forced systems with very small driving frequency. *Int. J. Bif. Chaos*, 14:999–1016, 2004.

A. Venkatesan and M. Lakshmanan. Nonlinear dynamics of damped and driven velocity-dependent systems. *Phys. Rev. E*, 55(5):5134–5146, 1997.

A. Venkatesan and M. Lakshmanan. Different routes to chaos via strange nonchaotic attractors in a quasiperiodically forced system. *Phys. Rev. E*, 58(3):3008–3016, 1998.

A. Venkatesan and M. Lakshmanan. Interruption of torus doubling bifurcation and genesis of strange nonchaotic attractors in a quasiperiodically forced map: Mechanisms and their characterizations. *Phys. Rev. E*, 63: 026219, 2001.

A. Venkatesan, M. Lakshmanan, A. Prasad, and R. Ramaswamy. Intermittency transitions to strange nonchaotic attractors in a quasiperiodically driven Duffing oscillator. *Phys. Rev. E*, 61(4):3641–3651, 2000.

A. Venkatesan, K. Murali, and M. Lakshmanan. Birth of strange nonchaotic attractors through type III intermittency. *Phys. Lett. A*, 259:246–253, 1999.

E. B. Vul, Ya G. Sinai, and K. M. Khanin. Feigenbaum universality and the thermodynamic formalism. *Russ. Math. Surv.*, 39:1–40, 1984.

A. Witt, U. Feudel, and A. Pikovsky. Birth of strange nonchaotic attractors due to interior crisis. *Physica D*, 109(1-2):180–190, 1997.

K. Yagasaki. Homoclinic tangles, phase locking, and chaos in a two-frequency perturbation of Duffing's equation. *J. Nonlinear Sci.*, 9:131–148, 1999.

T. Yalcinkaya and Y.-Ch. Lai. Blowout bifurcation route to strange nonchaotic attractors. *Phys. Rev. Lett.*, 77(25):5039–5042, 1996.

T. Yalcinkaya and Y.-Ch. Lai. Bifurcation to strange nonchaotic attractors. *Phys. Rev. E*, 56(2):1623–1630, 1997.

H. L. Yang. Milnor strange nonchaotic attractor with complex basin of attraction. *Phys. Rev. E*, 63:036208, 2001.

T. Yang and K. Bilimgut. Experimental results of strange nonchaotic phenomenon in a second-order quasi-periodically forced electronic circuit. *Phys. Lett. A*, 236:494–504, 1997.

Y. H. Yu, D. S. Kang, and D. C. Kim. Nonchaotic attractor with a highly fluctuating finite-time Lyapunov exponent in a hybrid optical system. *Journal of the Korean Physical Society*, 34(6):497–501, 1999a.

Y. H. Yu, D. C. Kim, J. Y. Ryu, and S. R. Hong. Experimental study on the blowout bifurcation route to strange nonchaotic attractor. *Journal of the Korean Physical Society*, 34(2):130–134, 1999b.

K.-P. Zeyer, A. F. Miinster, and F. W. Schneider. Quasiperiodic forcing of a chemical reaction: Experiments and calculations. *J. Phys. Chem.*, 99: 13173–13180, 1995.

T. Zhou, F. Moss, and A. Bulsara. Observation of a strange nonchaotic attractor in a multistable potential. *Phys. Rev. A*, 45(8):5394–5400, 1992.

Index

Arnold tongue, 107
Arnold's cat map, 72
autocorrelation function, 60

basin boundary bifurcation, 93, 95
bifurcation
  blowout, 113
  boundary crisis, 91
  interior crisis, 87
  Neimark-Sacker, 77
  phase-locking, 104
  pitchfork, 77, 111
    modulated, 26
    non-smooth, 111
  saddle-node, 77
  tori collision, 104
    non-smooth, 81
  torus doubling, 77
  transcritical, 77
blowout transition, 113
butterfly effect, 45

chaos, 5
Chebyshev polynomials, 169, 173
circle map, 14, 27, 43, 56, 137
  phase-locking, 104
  supercritical, 167
  tori collision, 104
codimension of criticality, 133
commutative subspace, 136
complexity, 1
continued fraction, 30

crisis
  boundary, 91, 93
  interior, 87, 93, 117
critical attractor, 160, 170
critical point, 132

dimension
  box-counting, 58, 96
  capacity, 58
  correlation, 58
  generalized, 58
  information, 58
  Lyapunov, 57
Diophantine number, 31
double crisis point, 93
driven pendulum, 191
Duffing oscillator, 23, 27

electronic circuit, 26
electronic oscillator, 187
excitable system, 27
experiments, 26

Feigenbaum–Kadanoff–Shenker equation, 143
Fibonacci numbers, 31, 134
final state sensitivity, 99
Floquet multiplier, 156, 167
fractal, 6, 57, 95
  basin boundary, 89, 93, 95
  torus, 63
fractional-linear map, 180, 182

Frenkel-Kontorova model, 27

golden mean, 10
  continued fraction, 31
GOPY model, 12
  autocorrelation function, 66
  blowout bifurcation, 114
  critical point, 137
  finite-time Lyapunov exponent, 52
  fractal dimension, 59
  modified, 14, 40
  phase sensitivity, 50
  power spectrum, 64, 67
  rational approximations, 33

Hénon map, 20, 43, 56
Harper equation, 179–180
Harper model, 20, 27

incommensurate frequencies, 2
intermittency, 119, 142, 172, 178, 184, 185
  crisis-induced, 90, 118
  on-off, 118

Josephson junction, 22, 27, 190
  current-voltage characteristics, 23

KAM theory, 29
Kaplan-Yorke formula, 57
Ketzmerick formula, 68

limit cycle, 1
logistic map, 16, 27
  boundary crisis, 91
  coupled, 114
  interior crisis, 87
  intermittency, 124
  rational approximations, 40
  torus doubling, 79
low frequency forcing, 56
Lyapunov chart, 162, 175
Lyapunov exponent, 5, 45, 46
  finite-time, 48–50, 126
  one-dimensional map, 11
  transverse, 113

magnetic ribbon, 26
Milnor attractor, 28
modulated pitchfork map, 12
multifractal, 58
multifractal formalism, 73
multifractality, 150

noise, 28

one-dimensional map, 11
overdamped pendulum, 22

pendulum, 27
phase sensitivity, 47
  exponent, 52
phase shift, 32
phase-locking, 107
Poincaré section, 30
power law, 150
power spectrum, 61

quasicrystal, 20, 179
quasiperiodic motion, 2
quasiperiodicity, 1

rational approximation, 29, 109, 125
renormalization group
  evolution operator, 134
  intermittency, 131
  period doubling, 131
  quasiperiodicity, 131
  relevant perturbation modes, 136
  scaling factors, 135
ring map, 43, 56
robustness, 4
rotation number, 14, 108, 137

scaling, 132
scaling coordinates, 145, 161, 163, 174
scaling factors, 133
Schrödinger equation, 27
Schwarzian derivative, 185
self-similarity, 133, 160, 161, 192
self-sustained oscillations, 1
sensitive dependence on initial conditions, 6, 45

silver mean, 31
singular continuous spectrum, 65, 73
  correlation dimension, 68
skew shift, 16, 27, 43, 59
SQUID, 26
stroboscopic
  map, 38
  method, 9
structural stability, 4
superstable cycle, 156
synchronization, 114

tori collision, 81, 104, 171, 176, 184, 191
  terminal point, 140
torus
  doubling, 79, 139
    cascade, 81
    terminal point, 81, 140
  fractalization, 83, 141
    point, 107, 123
  smooth, 33
transverse stability, 113
turbulence, 3

uncertainty exponent, 99
universality, 187
  class, 132, 135, 153, 190

Van der Pol oscillator, 27

Wiener-Khinchin relation, 62